Springer Tracts in Modern Physics
Volume 130

Managing Editor: G. Höhler
Editors: R. D. Peccei F. Steiner J. Trümper P. Wölfle

Honorary Editor: E. A. Niekisch

W0042320

Springer Tracts in Modern Physics

Volumes 100-119 are listed at the end of the book

---

*denotes a Volume which contains a Classified Index starting from Volume 36

Heinrich Stolz

Time-Resolved Light Scattering from Excitons

With 87 Figures

Springer-Verlag Berlin Heidelberg GmbH

Dr. Heinrich Stolz

Fachbereich Physik
Universität-Gesamthochschule Paderborn
Warburger Straße 100
D-33093 Paderborn

Physics and Astronomy Classification Scheme (PACS):
42.50.Md; 78.47.+p; 71.35.+z

ISBN 978-3-662-14909-6   ISBN 978-3-540-48395-3 (eBook)
DOI 10.1007/978-3-540-48395-3

Typesetting: Camera ready copy from the authors using a Springer TeX makro package
Production Editor: P. Treiber

SPIN: 10123321   56/3140 - 5 4 3 2 1 0 - Printed on acid-free paper

To M. K.

Preface

Since their theoretical introduction sixty years ago by *Frenkel, Peierls* and *Wannier*, excitons, the solid state analogue of hydrogen atoms, have played a central role in solid state physics. Recently, exciton physics has excited renewed interest due to the possibility of growing artificial semiconductor structures of reduced dimensionality, like quantum wells and dots. In these structures excitonic effects are dominant even at room temperature due to the strongly enhanced Coulomb interaction. Such structures allow one in principle to utilize the linear and nonlinear optical properties of exciton states in practical photonic devices.

Essential to all applications involving excitons is the understanding of the dynamical behaviour of the exciton states, i.e. the various relaxation processes that govern the decay of the exciton states following optical excitation. Here, one has to understand those processes that lead to a reduction of the occupation number of the states and give rise to *population* or *energy relaxation*. Even more important is the relaxation dynamics of the coherence of excitons, commonly called *phase relaxation*. This is established by excitation of the states with a well–defined phase, e.g. by a coherent laser pulse.

In contrast to atomic systems, where, by isolating the atomic species, the relaxation times can be made arbitrarily large, being limited only by radiative decay, the relaxation times of electronic excitations in solids are expected to be short because their interactions with other elementary excitations cannot be switched off. The investigation of exciton dynamics therefore requires appropriate ultrashort laser pulses and detection schemes. In recent years, nonlinear optical methods like photon echo and, more generally, four–wave mixing have been successfully applied to study the dynamics of excitons on a pico- and femtosecond time–scale.

However, since the pioneering work of the group of *Haken* it is well known that the nonlinear optical properties of excitons are determined by the deviation from their low density character as particles obeying Bose statistics. While this is an interesting topic by itself, it is highly desirable to investigate the coherent and incoherent dynamical properties of excitons by a linear optical method. In this work it will be shown that an elegant way to do this is provided by *time–resolved resonant light scattering* (TRLS) under transform-limited conditions, i.e. with simultaneously optimum temporal and spectral resolution. Essential to this goal is the realization of an experimental setup that allows one to reach the picosecond time range with single photon sensitivity. Such a setup is described in this book.

With this method it is possible to discriminate coherent contributions to the light scattering process, which can be identified as Raman scattering processes, from the incoherent hot luminescence. The coherent nature of the initially excited states can be directly made visible if the states are energetically split and temporal oscillations (*quantum beats*) occur in the TRLS. This quantum beating is demonstrated for different exciton systems including indirect excitons in silver bromide and exciton–polaritons in cuprous oxide. In all cases, a group theoretical analysis of exciton states and optical transitions is performed allowing one to deduce all relevant relaxation times from the experiments.

As another conceptually straightforward technique to measure the coherence of excited states *resonant Rayleigh scattering* is introduced. Here the elastically scattered light is emitted from states that have not experienced phase relaxation and, after resonant excitation by short laser pulses, decay with the coherence time of the states. It will be shown that for this type of scattering to occur in crystals, some kind of disorder is required to lift the quasimomentum selection rule. Excitons in gallium arsenide quantum well structures provide the first example for time–resolved resonance Rayleigh scattering.

With this volume, the author hopes to demonstrate the possibilities to study exciton dynamics in semiconductors by time–resolved resonant light scattering (TRLS).

Finally, I want to express my thanks to all those, whose support, encouragement and criticism made this work possible:
– W. von der Osten for the fruitful cooperation over the years, many enlightening discussions and for the opportunity to work in his group.
– All collaborators at the University of Paderborn, especially E. Schreiber, D. Schwarze, and V. Langer.
– S. Permogorov, Ioffe–Institute St. Petersburg, from whom I learned much about excitons and light scattering in semiconductors.
Thanks also go to G. Weimann, TU München, and D. Fröhlich, University of Dortmund for supplying the excellent samples that facilitated many of the investigations described in this book.
This work has been financially supported by the Deutschen Forschungsgemeinschaft and by the Minister für Wissenschaft und Forschung des Landes Nordrhein–Westfalen.

Paderborn, February 1994   *Heinrich Stolz*

Table of Contents

1. Introduction

Light scattering has long been known to provide a very powerful tool to investigate the interactions of light and matter. The basic ideas, which go back to *Rayleigh* and *Placzek* [1, 2], allow a simple picture of the light scattering process. Here a monochromatic electromagnetic wave with field strength $E_L(\omega_L)$ and frequency ω_L excites in the sample to be investigated a dielectric polarization according to the polarizibilty $\alpha(\omega)$:

$$E_L(\omega_L) \xrightarrow{\alpha(\omega_L, \omega_S)} \begin{array}{c} \text{oscillating} \\ \text{dipole} \end{array} \xrightarrow{} \begin{array}{c} \text{scattered light} \\ E_S(\omega_S) \end{array}$$

This polarization produces according to Maxwell's equations secondary waves that are superimposed on the primary excitation wave. In a stationary and homogeneous medium due to the extinction theorem by *Oseen* [3] only one propagating wave remains, having the same frequency as the primary wave but different wavevector and propagation velocity. All other waves are extinguished completely. Light scattering therefore occurs only if the extinction of the secondary waves is incomplete. This may occur for two reasons:

a) Spatial variations in the polarizibilty $\alpha(\omega_L, \omega_S)$. This kind of scattering is denoted as *Rayleigh scattering* because there is no frequency shift of the scattered light.

b) Temporal variation of the polarizibility $\alpha(\omega_L, \omega_S)$, e.g., by phonons. This gives rise to the usual *Raman processes*.

1.1 The Elementary Light Scattering Process

In the classical picture of light scattering the electromagnetic field and also the polarization of matter are treated as waves. However, as first discussed in the work of *Kramer* and *Heisenberg* (see e.g. [4]), light scattering is a quantum mechanical phenomenon that can be visualized in a particle picture as the absorption and emission of photons in a second order light–matter interaction process. In the process of spontaneous scattering discussed here, the emission of the scattered photon is induced by the vacuum fluctuations of

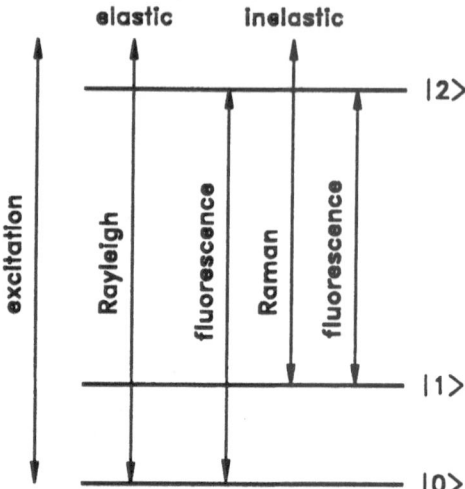

Fig. 1.1 Energy level scheme of a three–level system to indicate the fundamental processes in light scattering. After [5].

the electromagnetic field and is independent of the exciting light. Therefore, the scattered light is always superimposed on the spontaneous emission from the excited states of the system, all processes usually subsumed under the notion of *resonance fluorescence*.

The fundamental relations between these processes can already be discussed in the simple three–level system shown in Fig. 1.1 [4, 5]. The scattering process consists in the excitation of the system by destruction of the laser photon with energy E_L as indicated by the arrow marked 'excitation'. Due to energy conservation, for an isolated atom only two possible scattering processes exist:

- After scattering the atom returns to the ground state $|0\rangle$. This represents the elastic Rayleigh process with scattered photon energy $E_S = E_L$ and corresponds classically to the spatial inhomogeneous polarisability of a single atom.

- After scattering the atom remains in the excited state $|1\rangle$. This corresponds to inelastic Raman scattering with a scattered photon energy $E_S = E_L - (E_1 - E_0)$.

Both processes are most efficient for excitation energies E_L near the transition energy $E_2 - E_0$, i.e. in the case of *resonant scattering*.

A real atom, however, interacts with its surroundings. While in general this may be a thermal reservoir, even for an isolated system there is the interaction with the electromagnetic field giving rise to the radiative decay. Due to the interactions, the excited states have a finite lifetime leading to a spectral broadening of the states. Under resonant excitation the relaxation processes can provide for the difference between the excitation energy E_L and the transition energy resulting in a real occupation of the excited states. These then can decay into the ground state by spontaneous emission of the scattered photon. Therefore, resonant light scattering always occurs together with ordinary fluorescence, the balance betweeen the two being determined by the relaxation processes of the system. From these considerations it is obvious that the discrimation between proper light scattering and fluorescence is closely connected with the energy and phase relaxation of the states under resonant excitation.

For *stationary excitation* this topic has been discussed extensively in the literature (see e.g. [5 – 9]), whereby most investigations treat the more simple case of a two–level system. Accordingly, Rayleigh scattering and fluorescence can be discriminated by their different spectral properties. The spectrum of Rayleigh scattering coincides with that of the exciting laser, while the fluorescence lineshape is determined by the homogeneous linewidth of the excited state. The ratio between Rayleigh and fluorescence contributions is governed by the energy and phase relaxation times.

In contrast to the stationary case, for time–dependent scattering only a few studies exist. On the theoretical side, the early work of *Stenholm* [7], *Aihara* [10] and *Knöll* [11] has to be mentioned, while experimentally the study of light scattering from iodine molecules [12] was the pioneering work in this field. In these studies, however, the central criterion of a *time–resolved spectrum* where both the temporal and the spectral behaviour of the light is of equal importance was not realized. While for picosecond spectroscopic techniques based on nonlinear processes this problem is most often neglected —although it is of general nature due to the universality of the energy–time uncertainty—, for time–resolved light scattering it is of utmost importance. The importance of this question was first put forward by *Courtens* and *Szöke* [13] in their study of resonance fluorescence. At the same time *Eberly* stressed that any meaningful definition of a time–resolved spectrum requires a concrete experimental setup [14] that has to be taken into account properly in the theoretical description of resonance fluorescence [15, 16]. The theoretical treatment of *Khizniakov* [17, 18] included the investigation of resonance scattering from localized states, while later *Aihara* proposed a general theory of time–resolved scattering that includes even non–Markovian relaxation processes [19].

However, experimental investigations of the importance of the measurement process have not been performed systematically up to now and will be presented in this book. The experimental techniques providing simultaneously optimum spectral and temporal resolution necessary for this aim have been developed in recent years independently by *Saari* and *Freiberg* [20, 21] and in our group [22 – 25].

1.2 Light Scattering from Excitons:
A Historical Overview

It is now well established that the optical properties of almost all semiconductors and ionic crystals near the absorption edge are determined by excitons, at least at low temperatures. The investigations of these bound electron–hole pair states, which in many respects resemble a hydrogen atom in a solid, has thus been a major topic in solid state physics since the theoretical introduction of the exciton sixty years ago by *Frenkel, Peierls* and *Wannier* (see e.g. [26]).

The temporal dynamics of electronic excitations is most often studied by nonlinear optical methods like photon echo and —more generally— four–wave mixing ([27 – 34], for an overview of the method see [35]), mostly because of the possibility, with an appropriate excitation laser system, of reaching the femtosecond region. In this book the method of time–resolved light scattering (TRLS) is applied to study exciton dynamics, providing a linear method of investigations. Here it can be expected that this method will give new insights into the properties of excitons, in particular at low light intensities that normally cannot be realized by nonlinear spectroscopic techniques. Most important, however, is that light scattering almost naturally complies with the main difference between excitons and atomic systems, i.e., that the first are a spatially propagating, non–local electronic excitation of the matter system.

For stationary investigations of resonant light scattering processes in solids there exists a vast amount of literature that has revealed many important facets of the properties of exciton states (for an overview see e.g. [36]), the TRLS method was initiated by the need to study specific materials. On one hand, this was anthracene as a typical molecular crystal where the relevent states are Frenkel excitons that are restricted to a single molecule. This system was investigated by various groups, e.g. at the Academy of Science in Tartu (see [37, 38]). Here, however, the strong exciton–photon interaction leading to a marked polariton character of the states, hindered light scattering studies somewhat as only very thin layers can be used and multiple scattering has to be included. In this respect, systems with weak exciton–photon interactions are much easier to study, as is the case for the two materials cuprous oxide (Cu_2O) and silver bromide (AgBr) which interacts with light via a weak quadrupole transition (Cu_2O) or has an indirect band–to–band transition (AgBr). In the past, both materials have been already studied by stationary light scattering and as such have become model systems to investigate resonant light scattering from excitons (see e.g. [39 – 43]). In addition, the first time–resolved investigations of exciton relaxation were performed in these systems [22, 44 – 46]. In contrast, the more typical semiconductors from the III–V and II–VI groups could be investigated by TRLS only recently [47, 48] due to the improvements in time resolution.

The polarisation effects in the scattered intensities found in all these studies have again brought up the question about the nature of these resonant scattering processes, i.e. whether they represent Raman scattering or should rather be described as hot luminescence [49 – 51]. This problem always comes up if new types of light scattering processes are investigated, most recently in studying light scattering from excitons in quantum well structures (see [52, 53]), but could not be resolved satisfactorily up to now.

In this work, a systematic development of the light scattering will be given, including in particular the aspect of time resolution, to provide an answer to this problem. As the central point, a clear definition of what is to be understood by Raman scattering and how it is connected to the coherence of the intermediate states will emerge. This coherence can be measured in a precise way by *quantum beating* between the resonant states that show up in the light scattering process thus allowing a clear discrimination between the different contributions to the light scattering both in experiment and theory.

Thereby we will include, besides Raman scattering via phonons, also elastic light scattering comprising the Rayleigh process. To occur in a solid, this scattering requires the lifting of the quasi–momentum selction rule governing the exciton–photon interaction. These processes have long been considered as a disturbing 'dirt' effect, that sometimes renders the observation of low energy Raman processes impossible. Here we will demonstrate how its resonance characteristics and temporal behaviour can be used to explore in detail the relaxation dynamics of exciton states in disordered system.

1.3 Survey of the Book

The book starts with an overview of the general problem of time–resolved spectroscopy in Chap. 2 by introducing the *transform–limited time–resolved spectrum* as a way to characterize the temporal evolution of a light field. The fundamental quantity turns out to be the two–point field correlation function of the non–stationary scattered light that is convoluted with the spectral filter response to give the time–resolved spectrum. For coherent fields, it is shown that by the well-known two–dimensional phase retrieval problem the full time–dependence of the field in amplitude and phase can be recovered. The influence of the spectral filter is demonstrated for various examples. Concrete realizations of experimental arrangements are discussed with respect to their temporal response function, allowing one to give empirical design rules for these filters.

In Chap. 3 the state–of–the–art experimental setup realized by our group is presented. It allows transform–limited spectroscopy down to times of the order of 10 ps with single photon sensitivity.

A comprehensive treatment of the theory of exciton states and their relaxation processes by phonon scattering, as it is needed in the further chapters of this volume, is given in Chap. 4.

Chapter 5 is central for the theoretical discussion containing the treatment of the various light scattering processes. For the disorder–induced Rayleigh scattering a classical decription in the framework of Maxwell's theory is presented. This allows one to deduce the finite time response of the resonant Rayleigh process decaying with the coherence time of the electronic states.

To allow for depolarisation processes observed experimentally in the scattering, a density matrix treatment of light scattering is given that extends the previous description of light scattering by atomic systems [7 – 9] to degenerate states, typical for excitons in solids. This description, however, treats the scattering in the 'white–light' limit by neglecting the experimental setup and also ignores the dispersion of the excitons, i.e. the dependence of the exciton energy on the wavevector.

These effects can be systematically included in a quantum mechanical perturbation theory of light scattering that extends earlier calculations by *Aihara* [10]. The excitonic relaxation processes are modelled here as an interaction with a thermal reservoir in the Markov approximation. Raman processes can be included in this theory by an incoherent superposition of the different scattering channels corresponding to different wavevectors and phonon modes. Most important is that, due to the linear method of light scattering, the exciton states can be assumed to retain their boson character.

This theory is used to explain quantitatively the experimental results presented in Chaps. 6 – 8. Here time–resolved light scattering from a variety of quite different exciton states is discussed. These include

- Phonon scattering at the indirect exciton in AgBr,
- Phonon scattering at the quadrupole exciton–polariton in Cu_2O and
- Rayleigh scattering in GaAs/AlGaAs quantum well structures and ZnSe epitaxial layers.

The final chapter comprises some recent investigations not covered in the main text and discusses some future possibilities and trends of the TRLS method.

2. Time–Resolved Spectroscopy

2.1 Basic Considerations

2.1.1 The Definition of a Time–Resolved Spectrum

In order to develop the main principles of time–resolved spectroscopy the discussion has to start from a general point of view. We will assume the light to be described by a real quantity $V^{(r)}(\boldsymbol{r},t)$, which represents a component of the electric field strength or of the vector potential of the light field. Therefore any polarization effects will be neglected. While discussing the properties of light in a classical description, the formalism is put as close as possible to the quantum formalism by introducing the *complex analytic signal* [54] corresponding to the real field $V^{(r)}(\boldsymbol{r},t)$. This can be done by Fourier transforming $V^{(r)}(\boldsymbol{r},t)$ and noting that only the positive frequency components are of relevance since $V^{(r)}(\boldsymbol{r},t)$ is a real function. The analytic signal is then defined by

$$V(\boldsymbol{r},t) \equiv \frac{1}{2\pi} \int_0^\infty e^{-i\omega t} \left[\int_{-\infty}^\infty V^{(r)}(\boldsymbol{r},t') e^{i\omega t'} dt' \right] d\omega \ . \tag{2.1}$$

It is only composed of the Fourier components of the field strength with *positive* frequency and corresponds directly to the positive frequency part of the quantum mechanical field operator $E^{(+)}(\boldsymbol{r},t)$ [54].

For optical fields, the spatial and temporal dependence of the electric field strength is often assumed in the form of the *slowly varying amplitude and phase* (SVAP)–approximation

$$V(\boldsymbol{r},t) = V^{(0)}(\boldsymbol{r},t) \exp\left(i(\boldsymbol{k}_0 \cdot \boldsymbol{r} - \omega_0 t) + i\Phi(\boldsymbol{r},t)\right) \ . \tag{2.2}$$

Here $V^{(0)}(\boldsymbol{r},t)$ is a slowly varying amplitude and $\Phi(\boldsymbol{r},t)$ denotes the phase of the field. As even the shortest laser pulses produced up to now with a duration of 6 fs [55] contain several periods of oscillations, this approximation is valid for all light fields discussed here and will be taken for granted throughout this book[1].

---

[1] An obvious exception would be incoherent white light with vanishingly short coherence time. As we discuss only the effects of coherent light pulses, this is not of interest here.

To describe the physical situation realistically, the light field is represented by a stochastic variable. This means that it is a member of an ensemble of field strengths, $V(r,t)$ being a possible realization of the field occurring with a probability density $p(V)$. Obviously, V^0 and Φ are the corresponding stochastic variables of the SVAP approximation. Any experimentally measurable quantity given by the field–dependent function $f(V)$ can now be expressed as the ensemble average

$$\langle f(V) \rangle = \int f(V) p(V) dV \ . \qquad (2.3)$$

To measure the strength of electric fields in the optical spectral region only detectors are available that are sensitive to the intensity of the light field, defined as the average of the Poynting vector taken over several oscillation periods. This is in contrast to the situation in the microwave region, where field sensitive detectors, like fast Schottky diodes are common. Using specially designed photomultipler tubes, temporal resolutions of 20 ps with single–photon sensitivity are currently achievable (see Sect. 3.4), while with the fastest streak–cameras a temporal resolution of about 180 fs [56] is possible. Nonlinear mixing schemes, like up–conversion by sum–frequency mixing [57] give a time–resolution still better by a factor of two.

The analytic signal allows one to calculate in a simple way the *instantaneous intensity* of the light field, which is the Poynting vector averaged over one period of oscillation as

$$\langle I(r,t) \rangle = S_0 \langle V^*(r,t) V(r,t) \rangle \ . \qquad (2.4)$$

In the following the proportionality factor S_0 will be set equal to 1. This relation corresponds directly to the quantum mechanical expressions for the intensity [54].

Optical spectroscopy now aims to measure the strength of the different frequency components contained in a light signal, which represents then the *spectrum* of the light field. Obviously it is of no sense to directly decompose the field strength (Eq. (2.1)) itself, as in most cases the ensemble average will vanish. Instead the spectrum has to be defined in such a way that a spectral power density, i.e., an energy per time and frequency unit, is obtained. This requires the following assumptions to be made for a *physical* spectrum:

1. The spectrum should be a positive definite quantity.

2. The spectrum should be in principle measurable. This means that one should be able to construct a measuring apparatus, which, perhaps after some meaningful limiting procedure, gives the spectrum.

3. The spectrum should obey the causality principle. This means that it does not depend on quantities at later times than the time of measurement.

Since the work of Wiener it has been well known that the second order coherence function $G^{(1,1)}$

$$G^{(1,1)}(\boldsymbol{r}_1, t_1, \boldsymbol{r}_2, t_2) = \langle V^*(\boldsymbol{r}_1, t_1) V(\boldsymbol{r}_2, t_2) \rangle \tag{2.5}$$

allows one to define unambiguously the power spectrum in the case of a *stationary* light field [54]. The field correlation functions or coherence functions $G^{(m,n)}$ of order $N = m + n$ are defined for a light field $V(\boldsymbol{r}, t)$ by [54]

$$G^{(m,n)}(\boldsymbol{r}_1, t_1, \ldots, \boldsymbol{r}_m, t_m, \boldsymbol{r}_{m+1}, t_{m+1}, \ldots, \boldsymbol{r}_{m+n}, t_{m+n}) = \tag{2.6}$$
$$\langle V^*(\boldsymbol{r}_1, t_1) \ldots V^*(\boldsymbol{r}_m, t_m) V(\boldsymbol{r}_{m+1}, t_{m+1}) V(\boldsymbol{r}_{m+n}, t_{m+n}) \rangle$$

and allow one to characterize the properties of the field completely.

The second order correlation function $G^{(1,1)}$ of fields at the same position $(\boldsymbol{r}_1 = \boldsymbol{r}_2 = \boldsymbol{r})$ or the *autocorrelation function* will in the following be denoted by $G(\boldsymbol{r}, t_1, t_2))$. This has the fundamental symmetry property

$$G(\boldsymbol{r}, t_1, t_2) = G^*(\boldsymbol{r}, t_2, t_1) . \tag{2.7}$$

In the case of a stationary light field the autocorrelation function depends only on the difference of the time points $\tau = t_2 - t_1$ (note $t = t_2$)

$$G^{\text{stat}}(\boldsymbol{r}, \tau) = \langle V^*(\boldsymbol{r}, t - \tau) V(\boldsymbol{r}, t) \rangle . \tag{2.8}$$

The power spectrum $G(\omega)$ is then given by the Fourier transform of the autocorrelation function (Wiener–Khintchine theorem):

$$G^{\text{stat}}(\omega) = \int_{-\infty}^{\infty} G^{\text{stat}}(\boldsymbol{r}, \tau) e^{i\omega\tau} d\tau = 2\Re e \left\{ \int_0^{\infty} G^{\text{stat}}(\boldsymbol{r}, \tau) e^{i\omega\tau} d\tau \right\} . \tag{2.9}$$

Due to the analyticity of $V(\boldsymbol{r}, t)$ the power spectrum $G(\omega)$ does not contain negative frequency components and corresponds therefore to the physical spectrum that can be measured, e.g. by a frequency tunable spectral filter and a detector sensitive for the intensity of the light field.

Here we are interested in *non–stationary* optical processes, so that the above definition of the power spectrum is useless. In the past, there have been several attempts to extend the Wiener–Khintchine definition to time–dependent fields (for a review see [14]). The generally accepted definition was put forward by *Page* [58] and *Lampard* [59], and is used in recent papers [60]. In this approach one defines as frequency– and time–dependent spectrum

$$G_{\text{PL}}(\omega, t) \equiv \frac{\partial}{\partial t} \left\langle \left| \int_{-\infty}^{t} V(t') e^{i\omega t'} dt' \right|^2 \right\rangle . \tag{2.10}$$

In the following $G_{\text{PL}}(\omega, t)$ will be referred to as the PL spectrum.

While in a purely mathematical sense, this definition is obviously a valid extension of Eq. (2.9), as a derivative of a positive quantity it has the important drawback of becoming negative under certain circumstances. Furthermore it is shown below, that the PL spectrum is not even measurable with a physical setup. This excludes Eq. (2.10) as a general definition for the time–resolved spectrum of light. Other attempts to define a spectrum in an abstract way were also unsuccessful (see for details [14]).

Therefore it has been suggested by various authors (see [14,17–20]) not to relate the time–resolved spectrum to a formal mathematical definition, but instead to define it by the explicit use of a measuring apparatus. This setup has to consist of a spectral filter to select the frequency of the light field and a time resolving element to measure the temporal behaviour. The quantity that is measured with such a setup is called the *physical time–resolved spectrum of light* [14].

In the previous theoretical investigations a Fabry Perot etalon was considered as the spectral filter. Due to its fixed spectral bandwidth, which is determined by construction details, the practical use of this type of filter is very limited. Here we will extend *Eberly's* proposal to an arbitrary measurement setup and discuss in detail the requirements for spectral filters to be of use in time–resolved spectroscopy.

2.1.2 The Basic Measurement Setup

The basic setup to perform time–resolved spectroscopy is shown in Fig. 2.1. The material system under investigation is excited by a short laser pulse and emits light described as an electromagnetic field $E_S(t)$ from which we consider only one component $V(t)$ that is selected by a suitable polarization filter.

Quite generally, the temporal and spectral behaviour of this light field can be investigated in two ways. In the first (Fig. 2.1a) the light travels first through a spectral filter, which only transmits around a well–defined frequency ω_S. This filtered light is then detected by a time resolving photodetector. As time-resolved spectrum of this *direct procedure* one defines the intensity of the photodetector as a function of time and transmission frequency of the filter. In the second method (Fig. 2.1b), which usually is denoted as *sampling procedure*, the time resolution is determined by a temporal optical gate situated in front of the spectral filter. Accordingly, the time–resolved spectrum is the time–integrated intensity at the photodetector as function of sampling time and filter transmission frequency assuming that the integration time is large compared to the width of the optical gate.

Both methods of time–resolved spectroscopy require a finite amount of energy to fall onto the photodetector in order to register a signal. This implies for the direct method a finite spectral width of the filter, whereas for the sampling procedure this requires a finite width of the gate.

The time–resolved spectrum is obviously determined by the type of spectral filter used. Therefore one has first to consider the action of a spectral filter on the incoming electromagnetic field. There are basically two types of spectral filtering methods possible:

- Filters based on the interaction of radiation with matter that use, e.g. the absorption of light. This type of filter is most important for γ spectroscopy, like the Mößbauer effect [61] but will not discussed here further.

- Filters based on the interference of electromagnetic waves. Examples are the Fabry Perot interferometer and the diffraction grating. These filters are most important in the optical spectral region and will be discussed in more detail.

To be of use in optical spectroscopy any spectral filter should be *linear* and *causal*. Then the action of any spectral filter is given by a *temporal transmission function* $F_S(\omega_S, \tau)$ that depends on the mean transmitted frequency ω_S. This function represents the transmitted field strength for excitation with a $\delta(t)$–like pulse. The filtered electromagnetic field $V_D(t)$ is then given by

$$V_D(t) = \int_{-\infty}^{\infty} F_S(\omega_S, t - t') V(t') dt' . \tag{2.11}$$

For reasons of causality $F_S(\tau)$ must vanish identically for negative arguments $\tau < 0$.

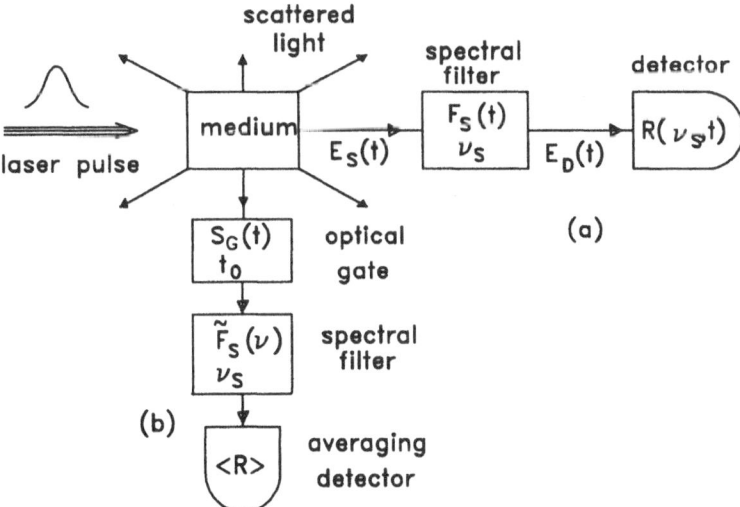

Fig. 2.1 Schematic setup for time–resolved spectroscopy. (**a**) direct method; (**b**) sampling method. For further explanations see text.

The action of a spectral filter can also be described in the frequency domain by the *spectral transmission function* $\tilde{F}_S(\omega, \omega_S)$ that is defined as the ratio of transmitted to incident field strength under monochromatic excitation. The transmission function can be parameterized by the mean transmission frequency ω_S and by the spectral width γ_S. Of course the exact functional relationship depends on the working principle of the filter.

A spectral filter will be called *bandwidth limited* if the spectral and temporal transmission functions are Fourier transforms of each other

$$F_S(\omega_S, t) = \frac{1}{2\pi} \int_{-\infty}^{\infty} \tilde{F}_S(\omega, \omega_S) e^{i\omega t} d\omega \ . \tag{2.12}$$

This is not necessarily the case, as the spectral or temporal behaviour of a filter often is determined by details of its construction, like slit width or spatial resolving power of the detector.

The filtered electric field impinges on a photodetector situated at position r_D resulting in a count rate $R(t)$. According to the quantum mechanical theory of light detection, this count rate is proportional to the instantaneous intensity given by Eq. (2.4) [5, 54].

In the temporal response of a real detector there are other additional broadening mechanisms, like transit time spreads of photoelectrons. These can be modeled by a intensity response function $D(t)$. Then the photon count signal is given by

$$R_D(\omega_S, t) = \int_{-\infty}^{\infty} D(t - t') \langle V_D^*(r, t') V_D(r, t') \rangle dt' \ . \tag{2.13}$$

This convolution shows how the information contained in the light field V_D is reduced by the finite time response.

Now time–resolved spectra have the maximum amount of information if

1. the spectral filter is bandwidth limited, and
2. the time response of the photodetector is short compared to that of the filter. In the following the former will be assumed to be $\delta(t)$–like to represent the ideal case.

The signal that is measured under these conditions as a function of time and filter frequency

$$R(\omega_S, t) = \int_{-\infty}^{\infty} dt' \int_{-\infty}^{\infty} dt'' F_S^*(\omega_S, t - t') F_S(\omega_S, t - t'') G(r, t', r, t'') \tag{2.14}$$

will be designated as the *transform–limited time–resolved spectrum* or as the *spectrochronogram* as is done by other authors [20].

Obviously, it is determined like the stationary Wiener–Khintchine spectrum by the second order coherence function, which, however, enters not via its Fourier transform, but as a more general linear functional. By definition, it fulfils the requirements of positive definiteness and measurability, but depends on the type of spectral filter used.

The influence of the spectral filter on the measured spectrum can be worked out more explicitly by defining the amplitude transmission function of the filter $F^{(0)}(\tau)$ according to

$$F_S(\tau) = F_S^{(0)}(\tau)e^{-i\omega_S\tau} . \tag{2.15}$$

Then Eq. (2.14) is written as

$$R(\omega_S, t) = 2\Re e \left\{ \int_0^\infty d\tau\, G_D(\mathbf{r}, t - \tau, \mathbf{r}, t)e^{i\omega_S\tau} \right\} . \tag{2.16}$$

Here the function $G_D(\mathbf{r}, t - \tau, \mathbf{r}, t)$ is given by

$$G_D(\mathbf{r}, t - \tau, \mathbf{r}, t) = \int_{-\infty}^\infty dt'\, F_S^{(0)*}(\omega_S, t - t' + \tau)F_S^{(0)}(\omega_S, t - t')G(\mathbf{r}, t' - \tau, \mathbf{r}, t') . \tag{2.17}$$

Now one can write down the time–resolved spectrum in close analogy to the power spectrum of stationary processes (Eq. (2.9)) as the Fourier transform of a 'coherence function' $G_D(\mathbf{r}, t - \tau, \mathbf{r}, t)$. However, this function has no direct physical meaning as a coherence function of a light field. Most important it is not the second order correlation function of the field at the photodetector. This is given by $\langle V_D^*(\mathbf{r}, t - \tau)V_D(\mathbf{r}, t)\rangle$, which is not identical to Eq. (2.17). Therefore, the relation (2.16) has only a formal mathematical sense.

Here one is tempted to try to define the time–resolved spectrum by the relation

$$R_{\mathrm{PL}}(\omega_S, t) = 2\Re e \int_0^\infty d\tau\, G(\mathbf{r}, t - \tau, \mathbf{r}, t)e^{i\omega_S\tau} \tag{2.18}$$

directly from the coherence function as has been done previously [60]. However, it can be shown easily that this definition is identical to the Page–Lampard spectrum Eq. (2.10).

By comparing Eqs. (2.17) and (2.18) the problem of measurability of the PL spectrum can be discussed, which due to the connection with the Wigner function [62] is important from a general point of view. To measure the PL spectrum the setup has to fulfil for all values of τ the following condition

$$G(\mathbf{r}, t - \tau, \mathbf{r}, t) = \int_{-\infty}^\infty dt'\, F_S^{(0)*}(\omega_S, t - t' + \tau)F_S^{(0)}(\omega_S, t - t')G(\mathbf{r}, t' - \tau, \mathbf{r}, t') . \tag{2.19}$$

This obviously is only possible, if the correlation function $G(\mathbf{r}, t - \tau, \mathbf{r}, t)$ does not depend on t, i.e. represents a stationary process. Then G can be put outside the integral, which implies that the filter function is constant. This is realized for a τ–independent $F_S^{(0)}$. Now for stationary processes it was shown in [14] that the PL spectrum is identical with the Wiener–Khintchine spectrum implying a trivial case for the measurability. In the non–stationary case therefore no experimental setup exists by which the PL spectrum can by measured.

In the same way it can by shown that the time–resolved spectrum associated with a concrete setup cannot be calculated from the PL spectrum directly. Instead one must always calculate the correlation function first. Therefore the use of the PL spectrum, as it is proposed in [60], seems to be of no use for practical measurement problems.

However, if the electromagnetic field to be measured is coherent, i.e., its second order correlation or coherence function, can be factorized according to [54]

$$G(\mathbf{r}, t', \mathbf{r}, t'') = V^*(\mathbf{r}', t')V(\mathbf{r}'', t'') , \qquad (2.20)$$

it is possible to retrieve the function $V(\mathbf{r}, t)$ unambiguously by introducing in Eq. (2.14) the Fourier transform of $F_S^{(0)}(t)$ given by $F_S^{(0)}(\Omega) = \int_\infty^\infty F^{(0)}(t)e^{-i\Omega t}d\Omega$. This results in the following expression

$$R(\omega_S, t) = \left| \int_\infty^\infty V_F(t', \Omega)e^{-i\omega_S t' + i\Omega t}dt'd\Omega \right|^2 \qquad (2.21)$$

with

$$V_F(t, \Omega) = V(t)F^{(0)}(\Omega)e^{-i\Omega t} . \qquad (2.22)$$

Therefore, to obtain $V_F(t, \Omega)$ and also $V(t)$ from the measured transform–limited spectrum $R(\omega_S, t)$ the two–dimensional phase retrieval problem has to be solved. For this it is well known that in two dimensions a unique solution exists (see e.g. [63]). As $V(t)$ can be obtained by a simple division from V_F, the problem of retrieval of the field strength by transform–limited spectroscopy is much simpler and, due to the redundancy implicit in the above relation, more tolerant to excess noise than other techniques, like the frequency resolved optical gating, that were proposed recently [64].

Using the *sampling procedure*, in addition to the dependence on the spectral filter, the time–resolved spectrum is determined by the optical gate. This can be characterized in a general way by the field amplitude gate function $S_G(t_0, T)$, parameterized by opening time t_0 and temporal width T thus giving the time resolution of the setup. Here the field at the detector is given by

$$V_D(t) = \int_{-\infty}^\infty F_S(t - t')V(t')S_G(t', t_0, T)dt' . \qquad (2.23)$$

The spectrum is the time–integrated counting rate of the detector

$$\overline{R}(\omega_S, t_0) = \int_{-\infty}^\infty dt \int_{-\infty}^\infty D(t - t')\langle V_D^*(\mathbf{r}, t')V_D(\mathbf{r}, t')\rangle dt' . \qquad (2.24)$$

This equation directly expresses the dependence of the time–resolved spectrum on the gate function and on the spectral filter resulting in a more indirect connection of the measured signal to the primary light field.

In the following, two idealized setups for transform–limited spectral filters will be discussed in order to gain a more profound understanding of the relevance of the filter for the time–resolved spectrum.

2.1.3 Transform–Limited Spectral Filters

Any spectral filter is characterized according to definition 2.12 by its temporal and spectral response functions. But only if these are connected via a Fourier transform relation, which has to be checked in each case, is the filter of use for transform–limited spectroscopy. Here we will discuss two types of spectral filters, which fulfil this criterion, the Fabry Perot etalon and the spectral slit function.

2.1.3.1 The Fabry Perot Etalon. As first example we consider a Fabry Perot etalon which was also used in earlier publications [14]. The filter action relies on the interference of the multiply reflected light beams as schematically shown in Fig. 2.2. The spectral transmission function is given be the Airy formula [3]:

$$\tilde{F}_S(\omega) = \frac{(1-r^2)e^{i\omega d/c}}{1-r^2 e^{2i\omega d/c}} \tag{2.25}$$

with d denoting the plate distance and r the amplitude reflection coefficient. If the finesse $F = \pi r/(1-r^2)$ is much larger than one, for frequencies around the n–th transmission maximum $\omega_S = n\pi c/d$ this relation can be simplified to

$$\tilde{F}_S(\omega) = \frac{i\gamma_S \cdot e^{i\omega d/c}/2}{\omega - \omega_S + i\gamma_S/2} \ . \tag{2.26}$$

Here $\gamma_S = (1-r^2)c/d$ denotes the full halfwidth of the transmittance function, which obviously has a simple Lorentzian profile. Due to the derivation of the Airy formula by phase conserving superposition of the electromagnetic fields the Fabry Perot etalon works transform–limited and its time response is given by Fourier transforming (Eq. 2.12) as

$$F_S(\omega_S, t) = \frac{\gamma_S}{2} e^{-\frac{\gamma_S}{2}(t-t_0)} e^{-i\omega_S(t-t_0)} \Theta(t-t_0) \ . \tag{2.27}$$

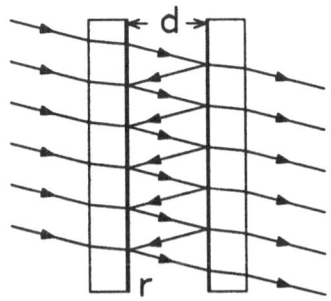

Fig. 2.2
Principle of a Fabry Perot etalon. d: distance of plane parallel plates with amplitude reflection coefficient r.

This is an exponential decay function with time constant $2/\gamma_S$ which is harmonically modulated with frequency ω_S. The Heavyside step function $\Theta(t-t_0)$ represents the retardation of the impulse response due to the finite propagation of light through the interferometer with $t_0 = d/c$ the transit time of the central ray. At first sight this result contradicts the picture of the temporal response of a Fabry Perot etalon to ultrashort laser pulses, i.e., a pulse train with a distance of twice t_0 decaying in amplitude. As Eq. (2.27) requires a very high finesse implying $r^2 \rightarrow 1$, this part of the transmission, however, has a negligible intensity. Such an etalon has virtually no transmission for very short pulses or for those with a very rapidly decaying autocorrelation function. Only those frequency components falling into the transmission maximum have finite intensity and these are just given by Eq. (2.27).

The relation between the temporal and spectral response of any spectral filter can be characterized comprehensively by an energy–time uncertainty relation, which is also denoted as the bandwidth product in laser theory. Here as bandwidth the full width at half maximum (FWHM) of the square of the amplitude transmission function will be taken. Then the bandwidth product of the Fabry Perot etalon is given by

$$\Delta\nu_{1/2}\cdot\Delta t_{1/2} = \frac{\ln 2}{2\pi} = 0.1103 \ . \tag{2.28}$$

2.1.3.2 The Spectral Slit Function. As the second example of a filter we will discuss the properties of the spectral slit function

$$\tilde{F}_S(\omega) = \begin{cases} e^{i\omega_S t_0} & |\omega - \omega_S| \leq \gamma_S/2 \\ 0 & |\omega - \omega_S| > \gamma_S/2 \end{cases} \tag{2.29}$$

Here t_0 is again the transit time of the electromagnetic field through the filter which is assumed to be the same for all transmitted components.

Assuming a transform–limited performance of this ideal spectral slit filter, which has to be checked for the concrete realization, the temporal response is given by

$$F_S(\omega_S, t) = \frac{2\sin(\gamma_S(t-t_0)/2)}{\gamma_S(t-t_0)}\cdot e^{-i\omega_S(t-t_0)}, \tag{2.30}$$

where the time retardation has been taken into account. Then for the spectral slit function the energy–time uncertainty relation is given by the well–known formula

$$\Delta\nu_{1/2}\cdot\Delta t_{1/2} = 0.886 \ . \tag{2.31}$$

The bandwidth product obviously is larger by a factor of 8 compared with the Fabry Perot etalon.

As is shown in Sect. 2.3.2, this kind of filter can be experimentally realized approximately with a subtractively mounted double monochromator. This type of filter has the big advantage that the spectral bandwidth can easily be

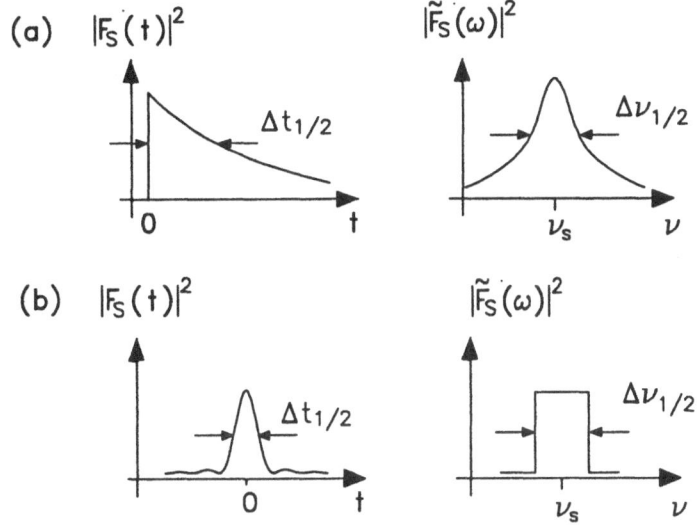

Fig. 2.3 Comparison between the temporal and spectral response of a Fabry Perot etalon (a) and of the ideal spectral slit (b). $\Delta\nu_{1/2}$ and $\Delta t_{1/2}$ are the FWHM of the square of the response function, $\nu = \omega/2\pi$.

varied by changing the width of a geometrical slit. It is therefore most suited for practical applications.

Looking more closely at Eq. (2.30) it turns out that the time response of the ideal slit filter covers all times and therefore does not obey the causality principle. As is shown by the discussion of concrete spectral filters in the next section, the temporal response has to be cut off in an appropriate way. For the subtractive double monochromator this is done by multiplying the temporal response function (Eq. 2.30) by a step function of the form

$$L(t) = \begin{cases} 1 - \frac{|t|}{T_0} & |t| \le T_0 \\ 0 & |t| > T_0 \end{cases} \qquad (2.32)$$

Here T_0 denotes the transit time difference of the extremal rays. The spectral response is then just the convolution of the slit function (Eq. (2.29) with the Fourier transform of Eq. (2.32)). However, for all practical purposes there is no difference in temporal and spectral behaviour as long as the relevant times are small compared to T_0.

By inserting into the double monochromator slits with a spatially varying transmission function other types of spectral filters can by realized. One possibility is a gaussian transmission profile

$$\tilde{F}_S(\omega) = e^{-\frac{(\omega - \omega_S)^2}{2\gamma_S^2}} \qquad (2.33)$$

which results in a temporal response function

$$F_S(\omega_S, t) = \frac{\gamma_S}{\sqrt{2\pi}} e^{-(\gamma_S t)^2/2} \cdot e^{-i\omega_S t} \ . \tag{2.34}$$

This filter has a bandwidth product of

$$\Delta\nu_{1/2} \cdot \Delta t_{1/2} = \frac{2}{\pi} = 0.637 \tag{2.35}$$

and rather smooth spectral and temporal cut–off characteristics.

For a spectral filter with triangular shape of width $\Delta\omega = \gamma_S$, that is approximately realized in a monochromator operating near the optimum spectral resolution, the temporal transmission function is given by

$$F_S(\omega_S, t) = \left[\frac{2\sin(\gamma_S t/2)}{\gamma_S t}\right]^2 \cdot e^{-i\omega_S t} \ . \tag{2.36}$$

In this case one obtains for the bandwidth product

$$\Delta\nu_{1/2} \cdot \Delta t_{1/2} = 0.319 \tag{2.37}$$

a much smaller value than for the ideal slit function.

The relations in the frequency and time domain valid for the Fabry Perot etalon and the ideal slit are graphically illustrated in Fig. 2.3. Obviously the Fabry Perot etalon has the smallest time–frequency uncertainty. This is closely related to the large extension of the Lorentzian filter function in the frequency domain compared to the sharp cut–off of the spectral slit.

2.2 Model Calculations for a Lorentz Oscillator

As an application of the foregoing discussion the time–resolved spectrum of a simple classical system emitting a time-dependent electromagnetic field, the damped Lorentzian oscillator, will be compared for the following experimental setups

1. direct measurement with a Fabry Perot etalon as spectral filter,
2. direct measurement with an ideal spectral slit,
3. measurement with optical sampling and Fabry Perot etalon as spectral filter.

For the first two cases the time response of the photodetector is assumed to be $\delta(t)$–function like.

The light field of a Lorentzian oscillator that is impulsively excited at $t = 0$ is given by

$$V(t) = V_0 \Theta(t) e^{-i(\omega_0 - i\gamma_0/2)t} \tag{2.38}$$

with lifetime $\tau = 1/\gamma_0$ and the Heavyside' step function $\Theta(t)$. The usual stationary spectrum of this oscillator is a Lorentzian with FWHM of γ_0. This spectrum formally can be gained by periodic excitation with a period long compared to the decay time and is normally that given in the literature.

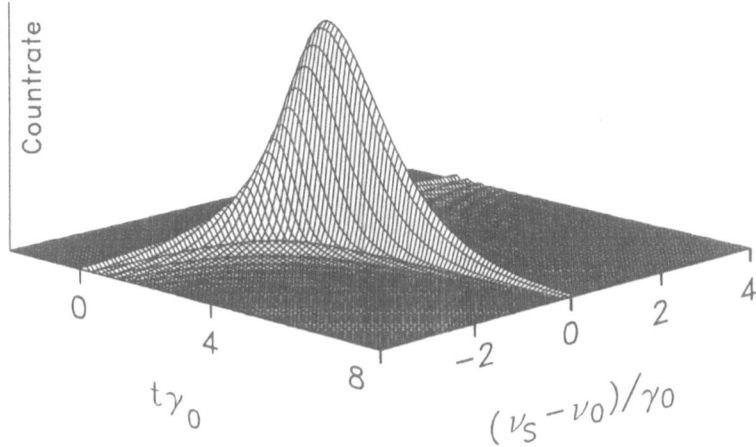

Fig. 2.4 Time–resolved spectrum of a Lorentz oscillator (time constant γ_0) measured with a Fabry Perot etalon as spectral filter. Plotted is the count rate versus time and versus difference of frequencies of the oscillator (ν_0) and the spectral filter (ν_S). Spectral bandwidth $\gamma_S = 1.5\gamma_0$.

2.2.1 Time–Resolved Spectrum with Fabry Perot Etalon

The calculation of the time–resolved spectrum according to Eq. (2.14) results in

$$R(\omega_S, t) = \frac{|V_0|^2 \cdot \gamma_S^2/4}{(\omega_0 - \omega_S)^2 + (\gamma_0 - \gamma_S)^2/4}$$

$$\times \left[e^{-\gamma_0 t} + e^{-\gamma_S t} - 2e^{-(\gamma_0+\gamma_S)t/2} \cdot \cos(\omega_0 - \omega_S)t \right] .$$

(2.39)

The characteristic feature of this time response is an interference term the frequency of which is given by the difference of oscillator and filter transmission frequency. This behaviour can be clearly seen in Fig. 2.4 where the time resolved spectrum is plotted for the case $\gamma_S = 1.5\gamma_0$. Most interesting is that at longer times the spectral response is narrower than the linewidth of the oscillator itself. This opens up the possibility to obtain with this type of filter spectral resolutions below the natural linewidth directly (see e.g. [65]).

Taking the limit of equal damping of filter and oscillator we obtain as time response

$$R(\omega_S, t) = |V_0|^2 \cdot \gamma_S^2/4 \cdot \left[\frac{\sin(\omega_0 - \omega_S)t/2}{(\omega_0 - \omega_S)/2} \right]^2 \cdot e^{-\gamma_0 t} . \qquad (2.40)$$

This is just the well–known temporal behaviour of the Mößbauer effect [61], which therefore can be viewed as one important example for transform–limited spectroscopy.

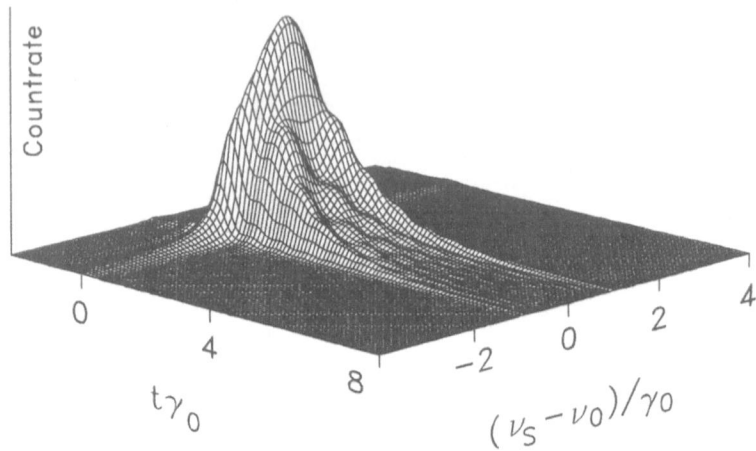

Fig. 2.5 Time–resolved spectrum of a Lorentz oscillator (time constant γ_0) measured with a spectral slit function. Plotted is the count rate versus time and versus difference of frequencies of the oscillator (ν_0) and the spectral filter (ν_S). Spectral bandwidth $\Delta\nu_{1/2} = 1.5\gamma_0$.

2.2.2 Time–Resolved Spectrum with Slit Function

Using the spectral slit function as filter (Eq. (2.29)) results in a completely different time–resolved spectrum. The field strength at the detector is given by

$$V_D(t) = \frac{V_0}{i\gamma_S} \cdot \left\{ \frac{1}{2} \ln\left(1 + \left(\frac{2\Delta\omega_+}{\gamma_0}\right)^2 \right) - \frac{1}{2}\ln\left(1 + \left(\frac{2\Delta\omega_-}{\gamma_0}\right)^2 \right) \right.$$
$$+ i\arctan\left(\frac{2\Delta\omega_+}{\gamma_0}\right) - i\arctan\left(\frac{2\Delta\omega_-}{\gamma_0}\right) \qquad (2.41)$$
$$\left. + \mathrm{Ein}\left(-(i\Delta\omega_- + \gamma_0/2)t \right) - \mathrm{Ein}\left(-(i\Delta\omega_+ + \gamma_0/2)t \right) \right\}$$

with $\omega_\pm = \omega_0 - \omega_S \pm \gamma_S/2$ and $\gamma_S = \Delta\nu_{1/2}$ the spectral width of the slit. $\mathrm{Ein}(z)$ denotes the modified exponential integral [66] defined by

$$\mathrm{Ein}(z) = -\sum_{n=1}^{\infty} \frac{(-1)^n z^n}{n!} \qquad |\arg z| < \pi . \qquad (2.42)$$

The time–resolved spectrum follows from $V_D(t)$ according to

$$R(\omega_S, t) = |V_D(t)|^2 \qquad (2.43)$$

This spectrum is plotted in Fig. 2.5 for the case $\gamma_S = 1.5\gamma_0$. Here we see pronounced temporal oscillations extending over the whole spectral range

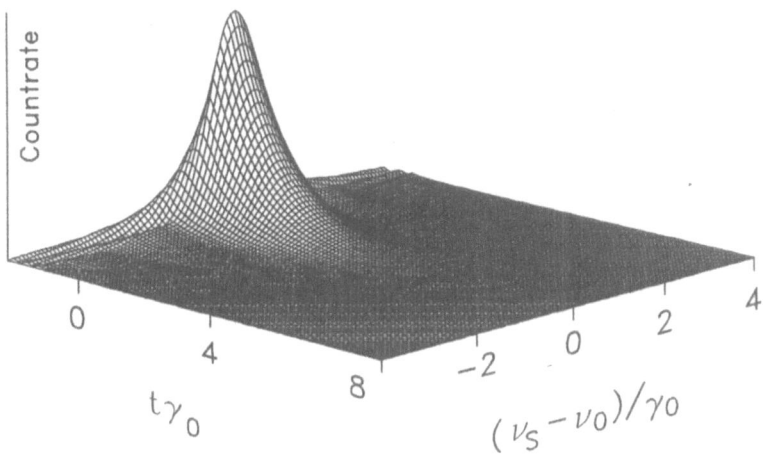

Fig. 2.6 Time–resolved spectrum of a Lorentz oscillator (time constant γ_0) measured with the sampling method using a Fabry Perot etalon as spectral filter. Plotted is the count rate versus time delay of the sampling gate (width $0.5/\gamma_0$) and versus difference of frequencies of the oscillator (ν_0) and the spectral filter (ν_S). Spectral bandwidth $\Delta\nu_{1/2} = 1.5\gamma_0$.

which exhibits a rather sharp cut–off. This reflects the spectral and temporal behaviour of the filter itself. In contrast to the Fabry Perot etalon no enhancement of spectral resolution by time discrimination is possible.

2.2.3 Time–Resolved Spectrum with Sampling Gate

To compare the sampling procedure with the direct time–resolved spectra we discuss the case of a rectangular time gate and a Fabry Perot etalon as spectral filter. The gate function is assumed to be

$$S(t, t_0, T) = \begin{cases} 1 & |t - t_0| \le T/2 \\ 0 & |t - t_0| > T/2 \end{cases} \tag{2.44}$$

Then Eq. (2.24) gives for the time–resolved spectrum

$$\overline{R}(\omega_S, t_0, T) \quad = \quad 0 \tag{2.45}$$

for $t_0 < -T$ and

$$\overline{R}(\omega_S, t_0, T) = \frac{|V_0|^2 \cdot \gamma_S^2/4}{(\omega_0 - \omega_S)^2 + (\gamma_0 - \gamma_S)^2/4} e^{-\gamma_0 t_0^*} \tag{2.46}$$

$$\times \left\{ \frac{1}{\gamma_0} \left(1 - e^{-\gamma_0 T^*} \right) \right.$$

$$+\frac{1}{\gamma_S}\left[1+e^{-\gamma_0 T^*}-2e^{-(\gamma_0+\gamma_S)T^*/2}\cdot\cos(\omega_0-\omega_S)T^*\right]$$

$$-\frac{\gamma_0+\gamma_S}{(\omega_0-\omega_S)^2+(\gamma_0+\gamma_S)^2/4}$$

$$\times\left[1-e^{-(\gamma_0+\gamma_S)T^*/2}\left(\cos(\omega_0-\omega_S)T^*-\frac{2(\omega_0-\omega_S)}{\gamma_0+\gamma_S}\sin(\omega_0-\omega_S)T^*\right)\right]\biggr\}$$

for $t_0>-T_0$. Here the following abbreviations have been introduced:

$$t_0^*=\begin{cases}0 & t_0<0\\ t_0 & t_0\geq 0\end{cases}\tag{2.47}$$

and

$$T^*=\begin{cases}t_0+T & t_0<0\\ T & t_0\geq 0\end{cases}.\tag{2.48}$$

The interference terms are also present, but as can be seen in Fig. 2.6 are restricted to small times comparable to the gate width. The spectral width of the time–resolved spectrum is mainly determined by the temporal width of the optical gate and therefore does not give any information about the physical process to be investigated. This is a serious drawback of this kind of time–resolved spectroscopy and should be considered in all practical applications.

2.3 Possible Experimental Realizations of Spectral Filters

The foregoing discussion has shown that for the experimental realization of the direct method in time–resolved spectroscopy the spectral filter is most important. In this section we therefore discuss several possibilities for transform–limited spectral filters and derive their temporal and spectral behaviour.

2.3.1 The Fabry Perot Etalon

The simplest spectral filter would be a Fabry Perot etalon. In order to be of practical use the free spectral range has to be larger than the spectral width of the signal to be investigated. For exciton states in solids this is of the order of 50 meV (or $\Delta\nu\simeq 10$ THz). According to Eq. (2.25) this implies a distance of the plates of some 10 μm. The value of the finesse is less critical, but determines the spectral resolution. This shows the main disadvantage of a Fabry Perot etalon as spectral filter! The bandwidth is fixed by construction and not variable. Except in some very special cases, this excludes the use of a Fabry Perot etalon in time-resolved spectroscopy.

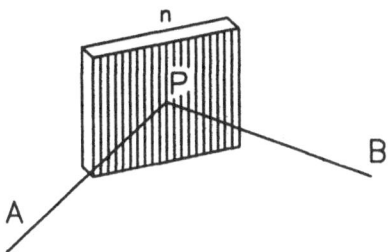

Fig. 2.7
Schematic diagram of optical rays for an arbitrary grating.

2.3.2 Grating Monochromators

2.3.2.1 General Properties. Much more important for practical applications are spectral filters that are based on diffraction gratings. In this section first some general results will be derived that are applicable to all kind of gratings. Then the temporal responses of different grating monochromators are discussed.

The temporal response of a single grating can be obtained from the geometrical grating theory using Fermat's principle. To this end we consider the optical imaging from point A to point B (see Fig. 2.7), which incorporates all elements of the optical setup including the diffraction grating. It is determined by minimizing the light path going through the pont P situated at the n–th grating rule according to a modified Fermat's principle [67]

$$L_{\text{eff}} = \overline{AP} + \overline{PB} + nm\lambda \ . \tag{2.49}$$

m denotes the grating order of the diffracted image. The last term of this relation takes into account the diffraction of light by the grating. The transit time of the light, however, is always given by the true path

$$L = \overline{AP} + \overline{PB} \ . \tag{2.50}$$

Comparing both relations, we immediately notice that a transit time difference occurs for rays through different points on the grating depending on the diffraction order m and the number N of illuminated grating rules

$$\Delta t = mN\lambda/c \ . \tag{2.51}$$

Due to the generality of the above derivation this relation is valid for all types of gratings, especially also for concave gratings, and for all types of mounting the gratings.

The transit time spread Δt is directly related to the diffraction–limited spectral resolution of the grating used in the filter. Taking not the Rayleigh criterion (resolution given by the first zero of the diffraction function [3]) but the FWHM of the square of the diffraction function, as is done here, the diffraction limited spectral resolution is given by

$$\frac{\Delta \nu_b}{\nu} = \frac{0.886}{mN} \ . \tag{2.52}$$

This shows that the transit time spread of the grating is connected by the enrgy–time uncertainty relation of the spectral slit

$$\Delta \nu_b \cdot \Delta t = 0.886 \tag{2.53}$$

with the diffraction limited resolution $\Delta \nu_b$.

For a more quantitative calculation we need the expression for the electromagnetic field strength $V_D(t)$ at point B in Fig. 2.7 that is emitted by a point source at A as the field $V(t)$. As usual in this section, we consider only light with well–defined polarization in a scalar theory. $V_D(t)$ is given by the superposition of all partial waves coming from different points P across the aperture of the grating in the following way:

$$V_D(B) = \int_{\text{aperture } \Omega} d\Omega(P)T(P)V(t - \tau(P)) \ . \tag{2.54}$$

Here $\tau(P)$ is the transit time of the light from A to B via P ($\tau(P) = L/c$) and $T(P)$ is the grating transmission function describing the changes in amplitude of the field at the grating in magnitude and phase.

Knowing the inverse function $P(\tau)$, the area integral can be transformed into an integral over τ. Comparing this to Eq. (2.11) the amplitude transmittance function $F_S(\tau)$ of the spectral filter can be obtained. For further calculation one needs of course the optical setup of the grating filter, no general solutions being possible. The case of some special grating monochromators is discussed in the next section.

2.3.2.2 Temporal Response. The simplest setup for a grating monochromator consist of a plane grating and two optical elements to focus the incident and scattered ray. In the Cerny–Turner mount (see Fig. 2.8), two concave mirrors are used for this purpose.

The properties of such a single grating monochromator as a spectral filter for time–resolved spectroscopy has been discussed in earlier work [68, 21]. Using the Czerny-Turner mount as an example, therefore, only the main results are discussed and compared to other filters.

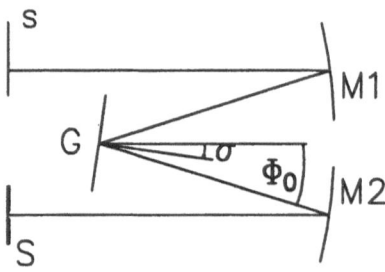

Fig. 2.8
Schematic of a grating monochromator in Czerny–Turner mount.
M1, M2 : focusing mirrors (focal length F), S: entrance slit, G: grating with constant d, s: spatial coordinate in the exit plane, Φ_0: Asymmetry angle.

Assuming as in Ref. [68] an aberration–free optical imaging, the transit time through the monochromator depends only on the position of point P given by the ruling number n. Then Eq. (2.54) can be evaluated quite straightforwardly, giving for the amplitude transmission function [68]:

$$F_S(\omega_S, t) = \frac{\cos(\sigma)}{T_0} f_m \exp(-i\omega_S(t - t_0))$$
$$\times (\Theta(t - t_0 + T_0/2) - \Theta(t - t_0 - T_0/2)) . \tag{2.55}$$

Here d is the grating constant, σ the diffraction angle at the transmission frequency ω_S in the m–th diffraction order. At this frequency the grating has an amplitude efficiency f_m which is obtained by a Fourier decomposition of the transmission function $T(P)$. $\Theta(t)$ denotes the Heavyside step function. The factor $\cos(\sigma)$ takes the reduction of the beam cross section into account occurring for a rotated grating position. t_0 denotes again the transit time of the central ray (see Fig. 2.8), which, for a typical high resolution monochromator with $F = 1$ m, amounts to about 10 ns.

The width of the temporal response $\Delta t_{1/2} = T_0$ is given by the transit time difference between the extremal rays

$$T_0 = \frac{2D}{c} \cos \Phi \sin \sigma . \tag{2.56}$$

The transmission frequency ω_S is determined by the rotation angle according to

$$\omega_S = \frac{\pi c m}{d \cos \Phi \sin \sigma} . \tag{2.57}$$

Going into the frequency domain Eq. (2.57) gives the usual diffraction limited spectral transmission function of a grating

$$\tilde{F}_S(\omega, \omega_S) = \cos \sigma e^{i\omega t_0} f_m \frac{\sin(T_0(\omega - \omega_S)/2)}{(T_0(\omega - \omega_S)/2)} e^{i(T_0(\omega - \omega_S)/2)} . \tag{2.58}$$

From relations (2.55) and (2.58) follows the diffraction limited bandwidth product of the single grating as

$$\Delta \nu_{1/2} \cdot \Delta t_{1/2} = 0.886 . \tag{2.59}$$

As expected, this time–frequency uncertainty relation agrees with the general expression (Eq. (2.53)).

However, Eq. (2.59) is only valid if the diffraction limited spectral resolution is realized in the monochromator. Now the usual setup for time–resolved fluorescence spectroscopy with picosecond temporal resolution consists of a single grating and a streak camera as photodetector, the photocathode of which is placed into the exit focal plane of the monochromator [69, 70]. This arrangement allows one to simultaneously resolve the temporal and spectral behaviour of the light field realizing the full multiplex advantage. But due to the limited spatial resolution of the streak tube, which is e.g. $30\mu m$ for a typical streak camera (type C1587 [71]), the intensity is integrated over a much larger frequency range than allowed by Eq. (2.59). Being rather insensitive to frequency, the temporal response is not changed. Therefore this setup does not work transform–limited under usual operating conditions.

To achieve transform–limited conditions with a single grating monochromator, according to Eq. (2.52) the number of illuminated rulings of the grating has to be reduced to such an extent that the diffraction limited resolution fulfils Eq. (2.53). Usually the spectral dispersion is given, e.g. by the spatial resolution of the photodetector. Therefore the illuminated area on the grating has to be reduced. Denoting by $\delta\omega_S/\delta s$ the linear dispersion of the monochromator, which is given by

$$\frac{\delta\omega_S}{\delta s} = \frac{\omega_S}{2F\tan\sigma}(1 - \tan\Phi\tan\sigma) \qquad (2.60)$$

with F the focal length of the Czerny–Turner mount. Assuming a spatial resolution of the detector of Δs_0, for which

$$\Delta s_0 = \frac{F}{D}\cdot\lambda_S\cdot 0.886\cdot\sqrt{1 - \sin^2\sigma}\ , \qquad (2.61)$$

the aperture ratio for the above mentioned streak camera can be calculated to be

$$\frac{F}{D} = 60\ ,$$

an extremely small value to which the entrance aperture has to be reduced. This is connected with a large reduction in intensity of the detected light which depends quadratically on F/D, and the multiplex advantage of the setup is completely destroyed.

A possible solution to this dilemma is given by Eq. (2.51). This shows that with a grating arrangement in zero diffractional order ($m = 0$) no transit time spread occurs. This can be realized, as is well known in optical spectroscopy, by using two identical gratings in the optical path, the first one working in order $m = +1$, the second one in order $m = -1$, resulting in an overall zero dispersion of the setup. This *subtractive grating mount*, first proposed by *Saari* [21] for use in time–resolved spectroscopy, requires a mirror–like symmetry of the two monochromators and therefore leads to identical pathways for the light beams.

Assembling two Czerny-Turner monochromators in the way displayed in Fig. 2.9, the usual subtractive double monochromator (SDM) is obtained that is also used in the experimental setup described in the next chapter.

The time response of this spectral filter can also be obtained similarly to Eq. (2.54). In the first step, the field $V(t)$ entering monochromator A in form of a point source in the center of slit S_1, has to be integrated over the aperture of A given by all points P_A at the grating 1. In this way one obtains the field in the focal plane of the middle slit S_2 as

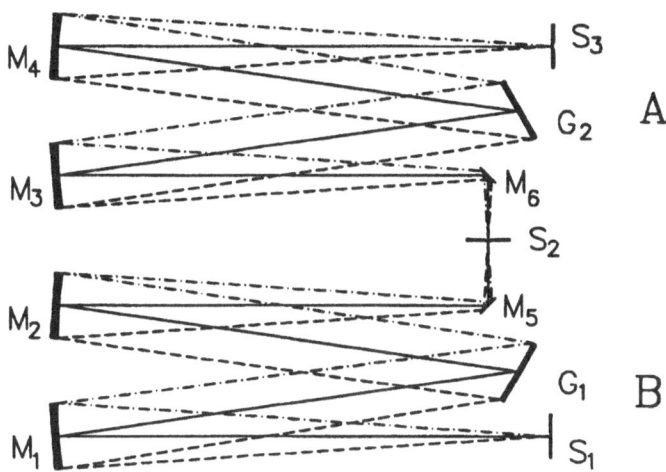

Fig. 2.9 Schematic diagram for a subtractive mounted Czerny-Turner double monochromator. G_1, G_2: grating, $M_1 \ldots M_4$: focusing mirrors, focal length F, $S_1 \ldots S_3$: slits, M_5, M_6: plane mirror. Full line: central ray, dashed and dashed-dotted lines: extremal rays. A and B are the two single monochromators.

$$V_A(t, s_A) = \int_{\text{aperture A}} d\Omega(P_A) T_A(P_A) V(t - \tau_A(P_A, s_A)) . \qquad (2.62)$$

To calculate the field at the exit slit S_3 of monochromator B (position of photodetector), in addition to the integration over the aperture of B (point P_B) one has to sum over all field strength $V_A(t, s_A)$ going through S_2. This gives finally for the detected field

$$V_D(s_B, t) = \int_{\text{slit } S_2} \int_{\text{aperture B}} ds_2 d\Omega_B(P_B) T_B(P_B) V_A(t - \tau_B(P_B, s_A)) . \quad (2.63)$$

This double convolution integral can be calculated by assuming an idealized geometry with exact mirror symmetry and by replacing the mirrors with aberration free lenses ('ideal' SDM). The result is (compare Eq. (2.30))

$$F_S(\omega_S, t) = \frac{2 \sin(\gamma_S(t - t_0)/2)}{\gamma_S(t - t_0)} \cdot e^{-i\omega_S(t - t_0)} \cdot L(t) , \qquad (2.64)$$

with $L(t)$ given by Eq. (2.32). For the time T_0 one obtains

$$T_0 = \frac{2D}{c} \cos \Phi \sin \sigma . \qquad (2.65)$$

The spectral bandwidth γ_S is determined by the width of the intermediate slit δs and can be calculated for a given grating from Eq. (2.60).

Using relation (2.64) the properties of the ideal SDM can be discussed. For large intermediate slitwidth the first factor is dominant giving the time response of the ideal spectral slit. Reducing the bandwidth, this goes over to the diffraction limited response of a single monochromator at very small slit widths. A similar conclusion was already drawn by *Saari* [21] from calculating numerically the propagation of special laser pulses in double monochromators.

At a sufficiently large intermediate slit width δs the time response of a SDM is only determined by δs. Most important, the response is independent of the width of the exit slit that can therefore be made arbitrarily small. Thus the time response does not depend on the spatial resolution of the photodetector also allowing the use of photomultiplier tubes with large photocathodes.

Furthermore, Eq. (2.64) predicts that the time resolution can be improved by increasing the width of the intermediate slit. However, experiments with very short laser pulses have shown that the time resolution as function of slit width has an optimum geting worse again for very large slit width. As this limits the practical applicability of the SDM as spectral filter, it is necessary to know the reasons for this behaviour. To this end, the time response of a real monochromator shown in Fig. 2.9 was calculated exactly taking into account the real behaviour of the optical elements.

To do this, in principle the double integral of Eq. (2.62) has to be performed. This integration, however, is only necessary if the spectral resolution is near diffraction limited, since then the entrance ray is diffracted at the intermediate slit losing its direction. In the case of much larger spectral resolution, in other words for a sufficiently large slit width, ray tracing based on geometrical optics is possible. The rays in monochromator B are then completely determined by those in monochromator A and the double integration is reduced to a simple one over the aperture of A. The transit times through both monochromators add to a total time τ which is determined only by the elongation θ and azimuth α of the rays and by the spectral coordinate δs at the intermediate slit, whereby we assume that a positive δs corresponds to an increase $\delta \lambda$ in wavelength.

Exact ray tracing performed for the Czerny-Turner mount shown in Fig. 2.9 now indicates that the assumption of equal transit times for all rays, that was the basis for the derivation of Eq. (2.64), is not exactly fulfilled. The difference in transit times between an arbitrary ray and the central ray can be expressed by the following semi–empirical relation

$$
\begin{aligned}
\delta T(\alpha, \theta, \lambda, \delta\lambda, \Phi_0) = {}& C_0(\Phi_0) \cdot F \cdot [\tan(\sigma(\lambda)) - 4/3 \cdot \Phi_0] \cdot \delta\lambda \\
& + C_1(\Phi_0) \cdot F \cdot \frac{\tan(\sigma(\lambda)) + \Phi_0}{\cos(\sigma(\lambda) - 2 \cdot \Phi_0)} \cdot \theta^2 \\
& + C_2(\Phi_0) \cdot F \cdot \frac{\tan(\sigma(\lambda)) - 4/3 \cdot \Phi_0}{\cos(\sigma(\lambda) - 2 \cdot \Phi_0)} \cdot \theta^3 \cdot \cos(\alpha) \\
& + C_3(\Phi_0) \cdot F \cdot \frac{\tan(\sigma(\lambda)) + 2/3 \cdot \Phi_0}{\cos(\sigma(\lambda) - 2 \cdot \Phi_0)} \cdot \theta^2 \cdot \cos(2\alpha) .
\end{aligned}
\tag{2.66}
$$

Here θ denotes the elongation of the ray, α its azimuth (all angles in radians) and F is the nominal focal length of the concave mirrors. σ gives the rotation angle of the grating at the wavelength $\lambda = 2\pi c/\omega_S$ (see Eq. (2.57)) and Φ_0 denotes the asymmetry angle of the Czerny-Turner mount. Finally $\delta\lambda$ gives the difference of the wavelength of the actual ray from the central wavelength λ. The constants C_i with $i = 0\ldots3$ are to be obtained by fitting this expression to the results of the numerical modeling. he results of this procedure for a typical asymmetry angle of $\Phi_0 = 4°$ are denoted in Table 2.1. The mean error of the transit times calculated with expression (2.66) amounts to about 1%.

According to Eq. (2.66) two effects are responsible for the degradation of the temporal response:

- the finite spectral bandwidth of the incoming light (first term)
- optical aberrations (all other terms).

Both effects scale linearly with the focal length of the monochromator. Even for a vanishing entrance aperture ($\theta = 0$) there remains a transit time spread. This is connected with the finite spectral extension of ultrashort light pulses. Increasing the spectral bandwidth of the filter, at first the temporal response shortens as expected from the time–frequency uncertainty. Then however, the transit time broadening given by the first term overwhelms and the time response gets worse reaching the order of 10 ps. Exactly this effect has been observed experimentally but could not be understood before.

To be more quantitative, Fig. 2.10 shows the calculated distribution of transit times for two different grating constants (given by the number of rules per mm) for the high resolution double monochromator used in the actual measurement setup (see next section) with opening ratio of $F/10$ and $F = 1$ m. Here we see that the useful time range is limited for gratings with 2400 lines/mm to about 5 ps. To achieve shorter time responses gratings with smaller number of rules have to be used. A reduction to 600 lines/mm results in a much better time response with 1 ps halfwidth, which is accompanied however by a reduction in spectral resolution by a factor of 4. The range where a grating with a specified number of rules can be used can be deduced from Fig. 2.11. This graph shows the dependence of the broadening t_B (given

Table 2.1   Parameters of the time response of a subtractive double monochromator in Czerny-Turner mount according to Eq. (2.66)

Φ_0	$C_0\,(\mathrm{ps/m\cdot nm})$	$C_1\,(\mathrm{ps/m\cdot rad^2})$	$C_2\,(\mathrm{ps/m\cdot rad^3})$	$C_3\,(\mathrm{ps/m\cdot rad^2})$
4°	−4.238	−307.114	−657.634	312.280

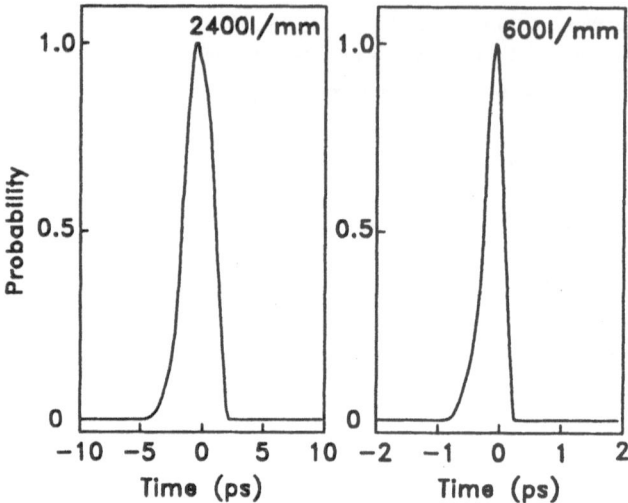

Fig. 2.10 Distribution of transit times of the subtractive double monochromator displayed in Fig. 2.9 for gratings with different lines per length unit.

as FWHM of the temporal response) on the pulsewidth of the incoming light pulses which are taken to be of a rectangular spectral shape with a transform–limited temporal duration given by Eq. (2.31). All other parameters are the same as for Fig.2.10.

Only for light pulses larger than the crossing point of the curve with the straight line $t_B = t_P$ can a grating be used as a transform–limited spectral filter. To reach even shorter times below 500 fs, which should be possible in the future by using improved streak cameras [56, 72] as photodetectors, the focal length of the monochromator has to be reduced without changing the aperture ratio.

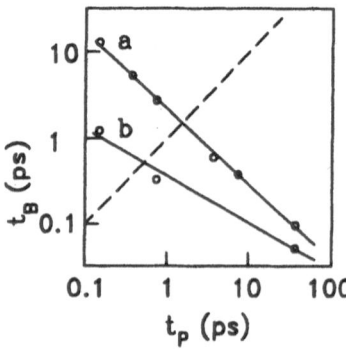

Fig. 2.11
Transit time spread t_B versus incoming optical pulse width t_P of a subtractive double monochromator ($F = 1$ m) for different gratings. (a: 2400 l/mm, b: 600 l/mm) in a double–logarithmic plot. Points: calculated results; full line: mean square fit. The dashed line represents the relation $t_B = t_P$.

3. Experimental Setup
for Transform–Limited Spectroscopy

The aim of this chapter is to describe in some detail the present status of experimental techniques to perform transform–limited time–resolved spectroscopy in the picosecond time range. This is done by presenting the setup which has been developed in our group during recent years [73–77]. This setup can of course also be used for more conventional time–resolved measurements.

The experimental setup consists mainly of two groups of devices, those to produce the ultrashort laser pulses and those to register the scattered light. The different components have to fulfill the following requirements:

- For generating the laser pulses

  - transform–limited light pulses, with negligible incoherent contributions and optimum relation between temporal duration and spectral bandwidth,
  - broad spectral region covering at least the visible range including the near ultraviolet and infrared,
  - low peak powers to avoid saturation effects, whereby high pulse repetition rates are possible,
  - variable pulse width ranging from 1 ps to 30 ps.

- For recording the scattered light

  - covering of the whole visible spectral domain,
  - large free spectral range allowing to work without prefiltering of the light,
  - variable spectral resolution,
  - detection of the scattered light with single photon sensitivity.

3.1 General Outline

The complete experimental setup is displayed schematically in Fig. 3.1. Besides the laser system and the components necessary for the spectral and temporal registration of the scattered light, it contains various devices, like an

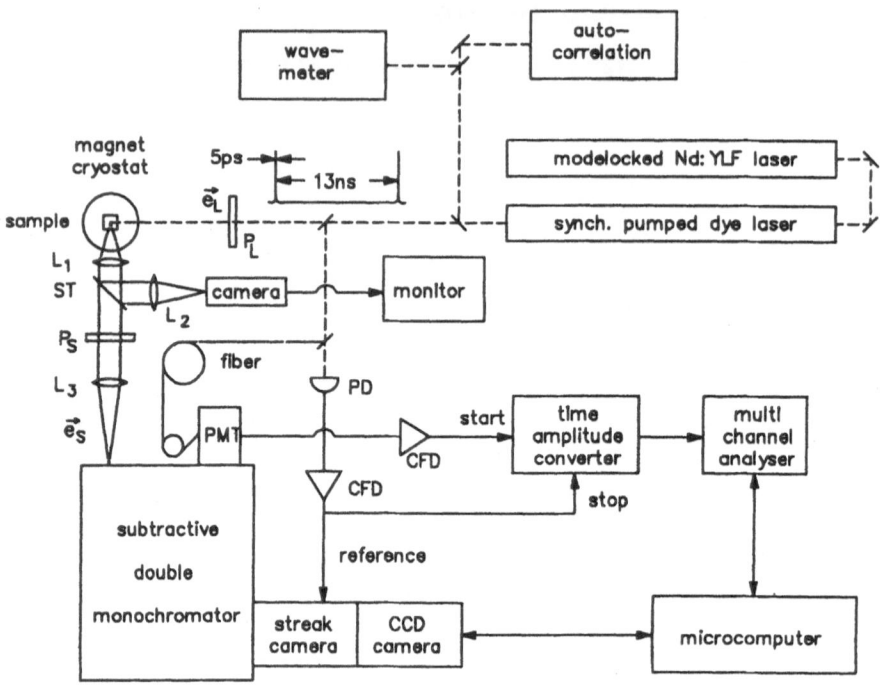

Fig. 3.1 Experimental setup for transform–limited spectroscopy. PMT microchannel plate photomultiplier, PD fast photodiode, CFD constant fraction discriminator. For further explanations, see text.

auto–correlator to measure the laser pulse width and a special monochromator to determine the laser wavelength, which allow one to control the proper operation of the whole apparatus. The important details of these components will be discussed in the next paragraphs.

Both the optical excitation of the sample and the detection of the scattered light requires a well–defined state of light polarization. Therefore the laser pulses traverse a plane polarizer followed by a polarization rotator and a $\lambda/4$ plate (denoted in Fig. 3.1 by P_L). The scattered light emitted from the sample is imaged by two lenses L_1 and L_3 onto the entrance slit of the monochromator. This arrangement provides a parallel beam part where the polarization state of the registered light is defined by a combination (P_S) of a plane polarizer (in combination with a $\lambda/4$ plate) and an achromatic $\lambda/2$ plate to select always the polarization direction with optimum transmission of the monochromator.

All experiments require the samples to be at low temperature ranging from 1.8 K (superfluid liquid helium) up to 100 K. Here different bath and variable temperature cryostats can be incorporated into the setup. For measurements in a magnetic field a superconducting magnet was used (type SM4, Oxford), producing fields up to $B = 4$ T.

Most important for practical measurements, one can directly image the sample onto a charge coupled device (CCD) TV camera via a beam splitter ST and a lens L_2 [74]. This allows one to observe the laser excited spot on the sample and to digitize the images with a commercial video digitizing system (Data Translation 2861)[76]. In this way a definite spot can be reproducibly excited and the laser beam quality checked.

The whole apparatus is controlled by a laboratory computer (HP Vectra RS/25C, type 100e, Hewlett–Packard) and a userfriendly menu–operated program system (for details see [76]). It supports the following types of measurements:

1. emission and excitation spectra,

2. time–resolved measurements with single–photon counting and synchroscan streak camera,

3. combination of these types of measurements to complex procedures.

3.2 Generation of Laser Pulses

The tunable excitation laser pulses are produced in a two–stage laser system consisting of a mode-locked Nd^+:YLF laser with a frequency doubler or tripler synchronously pumping a dye laser system. By using different laser dyes a spectral range from 375 nm to 850 nm can be covered.

3.2.1 Pump Laser

The pump laser that has emerged as being most useful for the purpose of time–resolved spectroscopy is a mode-locked Nd^+:YLF laser (Yttrium–Lithium–Fluoride doped with Nd^+ ions) (Model Antares, Coherent). The infrared output pulses are then frequency doubled in a KTP (potassium-tri-phospate) crystal and, to achieve ultraviolet output, mixed with the fundamental wave in a BBO (beta–barium–borate) crystal. The use of an argon ion laser is possible [73], but this does not provide enough ultraviolet power to efficiently pump blue laser dyes like stilbene 3 or coumarine 102 and use a saturable absorber in order to reach very short laser pulses. In the future, however, new laser systems like the Ti:sapphire laser [78] may provide attractive alternative light sources, also for this type of ultrashort spectroscopy.

The schematic setup of the pump laser is shown in Fig. 3.2. The Nd^+:YLF rod is pumped from two krypton arc lamps which together with the folded resonator mounted on an Invar bar provide good stability of the system. The mode coupling is done actively by an acousto–optic modulator (ML) working with an acoustic frequency of 38 MHz resulting in a laser pulse repetition rate

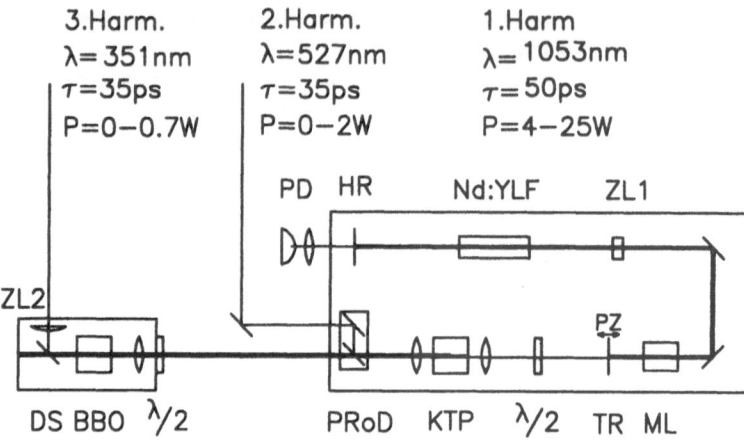

Fig. 3.2 Schematic diagram of the pump laser. ZL1, ZL2: cylinder lenses; HR, TR: high reflector, output coupler; PZ: piezo–element; ProD: polarization rotation and selection to obtain vertically polarized 523 nm pulses; PD: photodiode. For other explanations, see text. After [75, 79].

of 76 MHz. This corresponds to a temporal distance between two consecutive pulses of about 13 ns, a value to which the whole setup is adjusted. The stability of the commercial system turned out not to be sufficient for an optimum operation, as even changes of the cavity length of the order of 0.1 μm due to unavoidable drifts of the ambient temperature are large enough to degrade the pulses considerably. This problem could be overcome by a servo-loop adjusting the length of the resonator by means of a piezo-element changing the position of the output coupler mirror, whereby the error signal was derived either from minimizing the noise of the fundamental wave or maximizing the power of the second and third harmonic [80]. In this way pulse widths of the order of 35 ps could be obtained showing an almost gaussian time dependence. Typical output powers of the laser systems are 22 W at 1054 nm, 3 W at 527 nm and 1 W at 351 nm, enough to pump most dyes.

3.2.2 Dye Laser

The dye laser realized by us [75] combines the method of synchronous pumping and passive mode-locking by a saturable absorber to produce short, transform–limited laser pulses tunable over a wide spectral range (for a detailed discussion of this topic see e.g. [81]). To achieve proper transform–limited pulses in the blue spectral region, it is most important to have as few optical components in the resonator as possible, in order to minimize the group velocity dispersion. Here the six mirror cavity shown in Fig. 3.3 is the optimum arrangement. It consists of two pairs of concave mirrors to

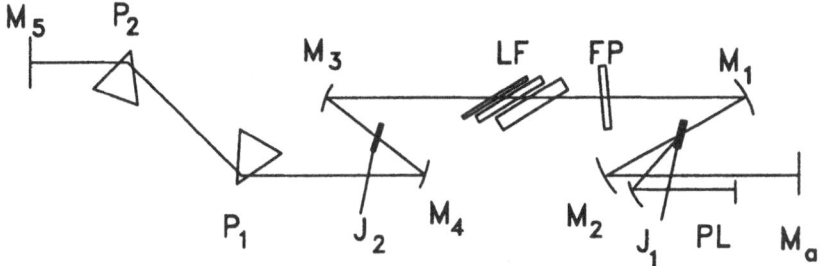

Fig. 3.3 Optical layout of the synchronously pumped dye laser. $M_1, \ldots, M_5, M_a$: mirrors; P_1, P_2: 60°–prisms for dispersion compensation; LF: Lyot filter; FP: etalon; J_1: pump jet; J_2: absorber jet; PL: pump laser beam. After [75].

provide the pump and absorber focal points. The dispersion compensation, which in the blue was already found to be necessary for pulses shorter than 5 ps, was achieved by a single prism pair that is doubly traversed during a cavity round–trip. The laser pulses therefore can be coupled out only at the other side of the resonator. One disadvantage of any configuration of this kind is that two laser pulses oscillate inside the resonator starting from the pump focus in opposite directions. By positioning the pump focus as near as possible to the output coupler and by a careful adjustment of the cavity optics one of the pulses could be suppressed completely. Some parameters of this laser system achieved up to now in the different spectral ranges are given in Table 3.1.

Table 3.1 Output power of the dye laser for different spectral regions. C102: coumarine 102; R6G: rhodamine 6G

dye	stilbene 3	C102	R6G	styryl 8
λ_{pump}	351 nm	351 nm	527 nm	527 nm
tuning range	425–465 nm	460–483 nm	570–630 nm	715–815 nm
output power (mW)	150	50	500	400
pump power (W)	0.65	0.65	2	2

3.2.3 Laser Pulse Diagnostics

To achieve optimum laser parameters both the temporal and spectral characteristics of the dye laser pulses are measured.

- Optical auto–correlator to measure the pulse width.
  The schematic optical layout of the pulse auto–correlator developed

in our group is shown in Fig. 3.4 [75, 82]. As translation element a magneto–dynamical shaker (Bruel&Kjaer) is used, which is driven by a frequency power generator allowing a wide range of variable scan velocities. As the spatial resolution of this system is much less than $\lambda/10$, interferometric correlation measurements can be performed with the setup easily [83]. To derive the correlation signal, the standard method of frequency doubling in a non–linear crystal can be used. Here the second harmonic is selected by appropriate interference filters. As another

method two–photon absorption of a fluorescence dye can by applied. This method has the advantage of a broad spectral response allowing one to characterize even very short pulses.

For a more routine monotoring of the pulse shape, a fast photodiode with a sampling oscilloscope or the synchroscan streak camera itself was employed.

- The spectral parameters of the laser pulses were measured with a specially designed grating monochromator shown in Fig. 3.5. The optical design is based on the Ebert mount using an echelle grating as dispersing element [77, 84, 85]. A CCD(charge–coupled–device) line array with 3456 pixels (Texas Instruments) served as photodetector. The wavelength of the laser to be measured is compared with the red line of a HeNe laser. The optics are designed in such a way that two orders fall at the same time onto the CCD detector allowing an absolute mea-

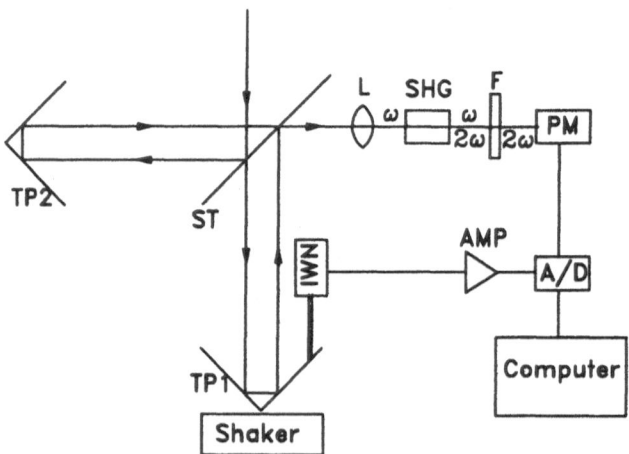

Fig. 3.4 Schematic setup of the pulse auto–correlator. Tp1,Tp2: triple prisms; IWN: inductive length transducer; SHG: frequency doubling crystal; ST: beam splitter; L: lens; F: interference filter; PMT: photomultiplier; A/D: analogue to digital converter. From [82].

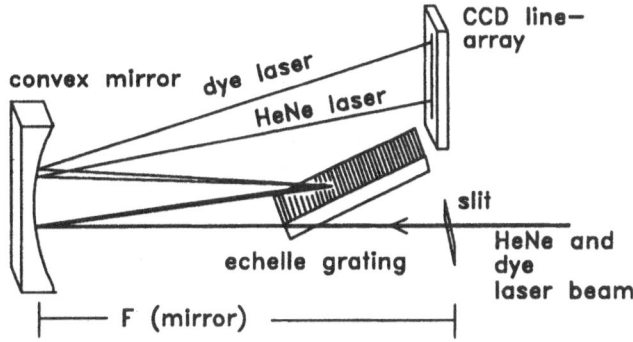

Fig. 3.5 Optical layout of the grating monochromator to measure the spectral parameters of the dye laser. From [85].

surement of the dye laser wavelength. The spectral resolution is about 0.005 nm with an absolute accuracy of 0.01 nm.

3.2.4 Typical Parameters

Most important for achieving optimum results in the measurement of time–resolved spectra is to have a sufficient variability of the duration and spectral width of the excitation laser pulses in order optimize the excitation of the resonant states. Provided the dispersion compensation is set properly, the pulsewidth depends only on the spectral width of the tuning element in the dye laser cavity. Using four different configurations, pulse widths from 400 fs up to 35 ps could be achieved. The shortest pulses with a duration of about 400 fs and 2 ps and corresponding spectral widths of 3.4 meV and 1.2 meV have been achieved by employing a one– and two–element birefringent filter, respectively. Here the requirements of the stability of the optical setup are most critical. For the application to time–resolved light scattering, pulses with a duration of about 5 ps have been found to be the best compromise between a small spectral width and short pulse duration. They are produced using a standard three–element birefringent filter giving a spectral width of about 0.4 meV. To obtain much longer and correspondingly spectrally very sharp laser pulses a single quartz etalon with a thickness of 1 mm is inserted into the resonator. Here extremely stable pulses with a halfwidth of 35 ps are produced.

The power spectra and auto–correlation traces for the different configurations are shown in Fig. 3.6 for stilbene 3 as an example. For the longer pulses the temporal shape could be measured by the streak camera. Here, in addition, the auto–correlation signal is shown in the inset of Fig. 3.6. The

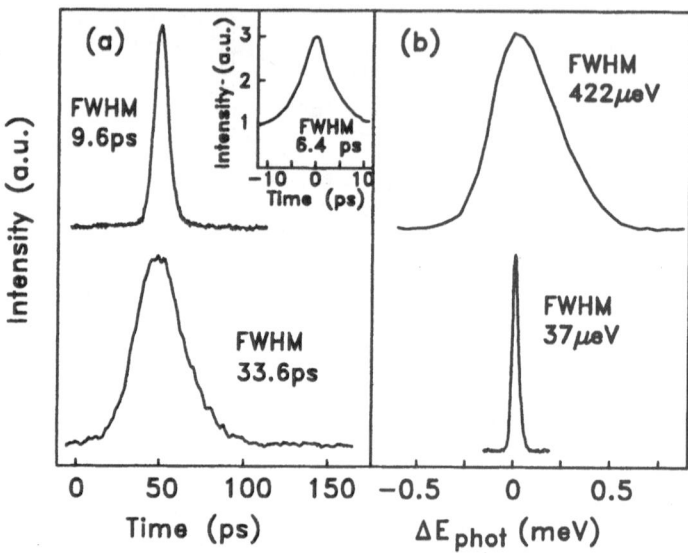

Fig. 3.6 Streak camera traces (a) and power spectra (b) for the dye laser equipped with a 3–element birefringent filter (upper traces) and an additional etalon (lower traces). Dye: stilbene 3. The inset shows the auto–correlation traces for the 3–element birefringent filter. Dye: styryl 8. The contrast ratio of 3:1 indicates optimum mode locking. From [75].

time–bandwidth products of all pulses are of the order of $\delta t \cdot \delta \nu = 0.3$, assuming a sech–like pulse shape [75].

3.3 The Spectral Filter

According to the discussion in the preceding chapter a subtractive double monochromator was employed to spectrally disperse the scattered light. This allows one to maximize the optical throughput and provides easily variable spectral resolution while providing a high stray–light rejection of the order of 10^{-12} that is essential to measure weak scattering lines beside a strong Rayleigh scattered laser line. Two different models were at our disposal differing in optical layout

- Model B&M 100S (B&M–Spektronik), with a focal length of 1 m and equipped with holographic gratings for the blue and ultraviolet spectral range (2400 l/mm). It consists of two separate single monochromators coupled optically by two focusing mirrors and driven mechanically by a common lead screw [73].

- Model Jarrel–Ash 25–103, with gratings for the red spectral region (1200 l/mm). This monochromator has the advantage of the simple optical layout shown in Fig. 2.9 where the gratings are mounted on a common rotation axis providing high mechanical stability.

For the measurements described below we used in most cases the B&M type, which shows an additional temporal broadening of about 5 ps, a value somewhat larger than that expected from Fig. 2.11 due to the more complicated optical layout.

3.4 Light Detectors

In order to have the necessary sensitivity for light scattering studies, optical detectors based on the photoelectric effect have to be employed. In our setup we used for this purpose either a synchroscan streak camera (model C1587, Hamamatsu) or a microchannel–plate photomultiplier (model 2566U, Hamamatsu), working in the single–photon counting mode. For both detectors the typical temporal response to laser pulses of 5 ps duration is shown in Fig. 3.7. The time resolution that can be achieved is around 10 ps for the streak camera and 20 ps for the photomultiplier.

3.4.1 Streak Cameras

The principle of time measurement by a streak camera is the time–position conversion of an optical pulse by deflecting the photoelectrons emitted from a photocathode into an electron optics similar to the common oscilloscope tube and accelerating the electrons towards a phosphor screen [70]. In our setup the streak camera has the two main purposes

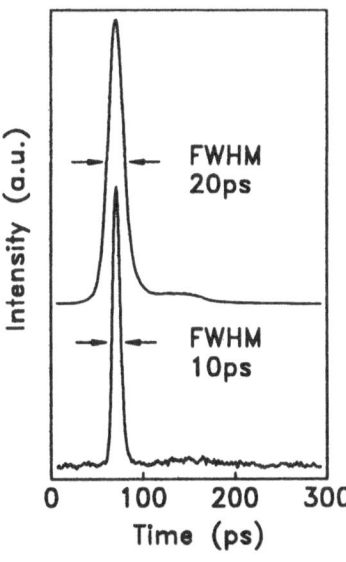

Fig. 3.7
Time response of the experimental setup to laser pulses of 5 ps duration (as obtained from the auto–correlation trace). Upper trace: single photon counting; lower curve: synchroscan streak camera.

- to check the quality, the temporal shape and time jitter of the laser pulses by direct coupling of the laser light through a quartz fibre onto the photocathode [86] providing in addition a definite zero–time reference,

- to detect the scattered light at the exit slit of the monochromator.

The streak camera operates in the setup in the synchroscan mode [70], in which the deflection is provided by a high–frequency voltage synchronous with the repetition frequency of the laser pulses. In order to achieve the highest time resolution this was derived directly from the signal of a fast photodiode detecting the dye laser pulse train. In this configuration an optimum time resolution of better than 7 ps FWHM is achievable, while in day–to–day use 10 ps are routinely possible.

To read out the streak signals a standard TV camera system (Grundig FA76 with Si–vidicon 4532) is optically coupled to the phosphor screen. The output of the camera is fed at a repetition rate of 25 Hz into a digitizing frame acquisition interface (Data Translation 2856) allowing fast image processing [76]. In this system the dynamics of the streak camera/readout system is limited by a bad signal–to–noise ratio to about two orders of magnitude (compare the lower trace in Fig. 3.7).

3.4.2 Single Photon Counting with Photomultipliers

A single–photon sensitivity and a high dynamic range of four orders of magnitude is possible by using photomultipliers (PMT) as photodetectors using the method of time–resolved (or time–correlated) single photon counting (see e.g. [87]).

The main component of this method of photon counting is a time–amplitude converter (TAC, see Fig. 3.1) that provides an output pulse with amplitude proportional to the difference between the start and stop pulse arrival time. In order to count every single scattered photon, it is used in the setup in an inverted mode, whereby the conversion is started by the PMT pulses and stopped by a laser pulse following next in time. The time window accessible is limited to the time between the laser pulses (13 ns). The output pulses of the TAC are digitized in a conventional multi–channel analyzer (resolution 14 bit) and are stored as a time profile.

While conventional PMT using dynodes to amplify the single photoelectrons have at best time-resolutions of the order of 200 ps, PMT working with microchannel–plates as amplification medium achieve much better resolution, both in the direct pulse output and in the single–photon mode. Using a sufficiently fast trigger electronics the time–resolution is limited by the transit-time spread of the electrons in the amplification cascade. Earlier types of microchannels had a channel width of $12\mu m$ and time–resolutions of 40 ps were possible. The latest types have a reduced channel width of $6\mu m$ allowing resolutions of below 20 ps. If the technology to further reduce the channel diameter is available, further improvements in time-resolution are possible.

To achieve this temporal resolution the electronic discriminators transforming the anode pulses from the PMT, which show pulse–height fluctuations due to the amplification process, into standard shaped pulses to trigger the TAC should have a wide bandwidth in order to respond properly on the fast PMT pulses. With a risetime of the order of 100 ps, bandwidths in excess of 3 GHz are necessary. Since commercial discriminators, however, have a bandwidth of about 500 MHz, we had to develop appropriate electronic circuits ourselves that are discussed briefly in the next section.

Despite a careful layout of the electronic setup using high–quality HF cables, connectors and amplifiers, due to changes in temperature a long–term drift of the time–zero is unavoidable. This drift is of the order of some ps during a day and may render impossible the analysis of the decay curves, in particular for long–time measurement runs and short decay times of the phenomena to be observed. Here the direct coupling of the excitation laser pulses using an optical single–mode quartz fibre on the photocathode of the PMT provides an accurate time reference without disturbing the measurements [74].

3.4.3 Constant Fraction Discriminators

The development of broadband electronic discriminators employing microwave techniques with multi–GHz bandwidth using the *constant-fraction* principle was essential to achieve time–resolutions in the picosecond range with photomultiplier tubes [89].

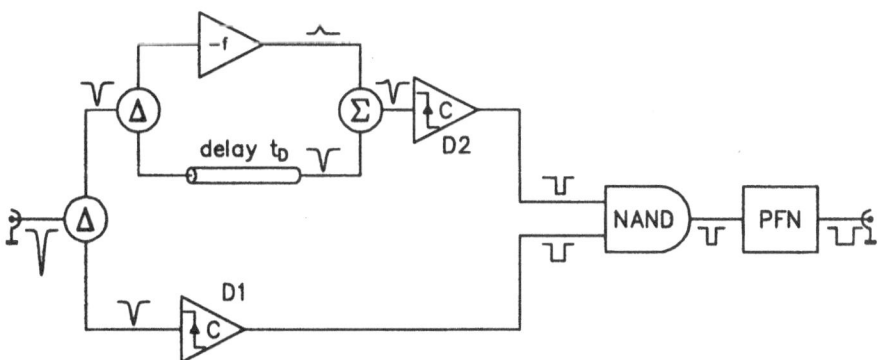

Fig. 3.8 Schematic layout of a constant fraction discriminator to obtain a standard–shaped electrical pulse from the PMT single photon pulse using microwave techniques. Δ: power splitter; Σ: power combiner; D1,D2: ultrafast comparator; PFN: pulse forming network. After [89].

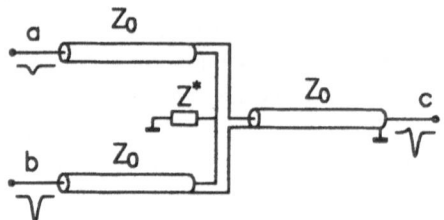

Fig. 3.9
Microwave circuit of the 180° power combiner.
Z_0: transmission line impedance;
Z^*: matching impedance. After [89].

The principle electronic layout of a constant fraction discriminator (CFD) is depicted in Fig. 3.8. The input pulse, the amplitude of which is varying statistically, is split into two pulses. The first pulse triggers a conventional amplitude comparator D1, so that only pulses with amplitudes exceeding a certain threshold are transmitted. The second pulse goes through a second power splitter resulting in two pulses of the same amplitude. The first of these is inverted and damped by the factor f. The other is delayed by the time t_D. Both are then summed in a power combiner forming the constant–fraction pulse. By choosing the damping factor f and the delay properly, the zero–crossing of the constant–fraction signal that triggers the comparator D2 can be made independent of the pulse height [89]. A valid output–pulse is obtained by the NAND gate only if the pulse height of the input pulse exceeds the trigger level of the amplitude comparator, the time of the output pulse being determined by the zero–crossing of the CFD. The critical parts that require a large bandwidth are the power splitter and combiner. While commercially available CFDs use active electronic devices limiting the bandwidth to about 400 MHz [88], the use of the passive microwave–network shown in Fig. 3.9 gives a bandwidth in excess of 4 GHz. As comparators, high sensitivity devices with a gain–bandwidth product of 600 GHz (type VC7697, VTC) are employed.

3.5 Transform–Limited Properties of the Setup

As shown by Fig. 3.7 the time–resolution is presently limited by the time response of the photodetectors to 10 ps. As the temporal broadening of the double monochromator of 5 ps is below this value, transform–limited detection of the scattered light should be possible down to times of 10 ps. To demonstrate the actual possibilities of the setup the temporal response of the system to laser pulses of 5 ps duration falling directly onto the entrance slit of the monochromator was investigated as a function of the spectral bandwidth using the streak camera as detector. As dye styryl 8 was used and the double monochromator (B&M 100S) was equipped with gratings having a ruling of 1200 lines/mm. In this case the residual time–spread is about 3 ps and does not influence the results.

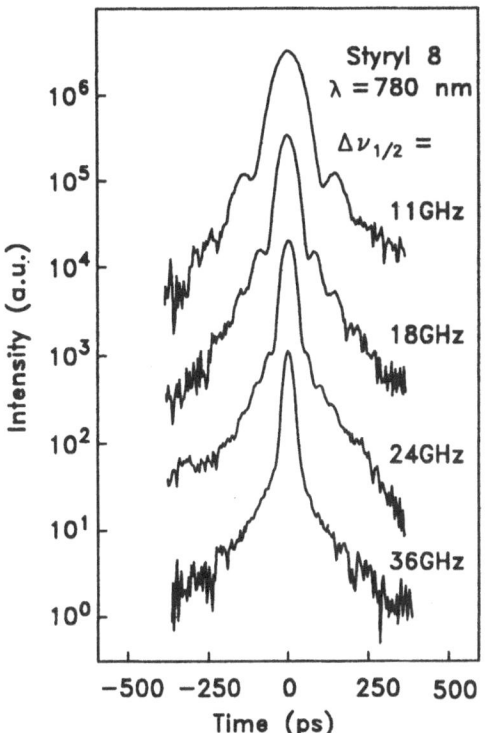

Fig. 3.10
Time response of the setup in Fig. 3.1 to 5 ps laser pulses for various spectral bandwidths $\Delta\nu_{1/2}$ of the monochromator using the synchroscan streak camera as detector. Note the logarithmic intensity scale.

The results of this experiment are shown in Fig. 3.10 and clearly show the expected dependence of the time response on the spectral bandwidth. For large values (curve for 36 GHz) the time resolution is determined only by the streak camera itself (in this experiment about 20 ps). Reducing the spectral bandwidth, the width of the temporal response function gets larger and side-maxima show up, which correspond to the $\sin t/t$ response of the rectangular slit function (compare Fig. 2.3), since for short, $\delta(t)$–like excitation Eq. (2.30) is just the limit of the relation (2.41). It should be noted that even for quite large spectral bandwidth the side wings of the temporal response are still influenced by the transform–limitation. This shows that for a quantitative analysis of time–resolved spectra the temporal behaviour of the spectral filter has to be taken into account.

The quantitative analysis of the temporal halfwidth of the curves in Fig. 3.10 gives the data which are plotted as full dots in Fig. 3.11 versus the reciprocal of the corresponding spectral width. By subtracting, under the assumption of statistical independence, the finite temporal resolution of the streak camera, which is given by the limit of $\Delta t_{1/2}$ for $1/\Delta\nu_{1/2} = 0$, the triangles are obtained. These fall nicely on the expected relation $\Delta\nu_{1/2}\cdot\Delta t_{1/2} = 0.886$ of the time–bandwidth product of a rectangular spectral filter, fully confirming the calculations from Chap. 2.

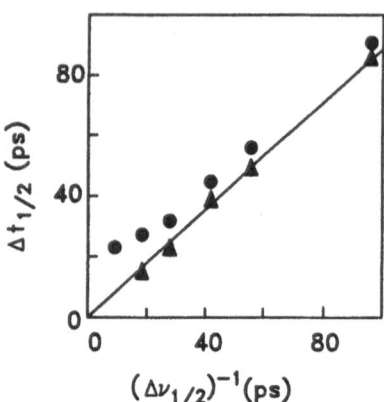

Fig. 3.11
Comparison of the temporal resolution $\Delta t_{1/2}$ with the reciprocal spectral width $\Delta \nu_{1/2}$. Full dots: experimental results; triangles: values obtained after subtraction of the temporal resolution of the streak camera as discussed in the text. The straight line represents the theoretically expected relation $\Delta \nu_{1/2} \cdot \Delta t_{1/2} = 0.886$ (Eq. (2.31)).

3.6 Scattering Geometries

As the setup will be used in solving quite diverse problems, a flexible scattering geometry with a variable configuration for the excitation and detection directions and the polarization directions is mandatory. For example measurements employing a 180° back–scattering geometry [90, 48], Brewster geometry (Chap. 8), 90° geometry (Chap. 6) and in a 0° transmission geometry (Chap. 7) are possible. In particular for the 180° and 0° configuration, the fluorescence light from the dye laser that is emitted in the same direction as the laser pulses due to ordinary fluorescence of the dye molecules has to be filtered out efficiently. For this purpose, a specially designed grating monochromator [91] can be inserted in the excitation beam; it suppresses the fluorescence light by a factor of 10^{-5} without broadening the laser pulses too much.

4. Exciton States in Semiconductors

In accordance with their importance for the optical properties of solids, excitons are a very lively subject of solid state physics and a number of excellent reviews on various topics exist (see e.g. [26, 92, 93]). Therefore only those aspects will be discussed here that are important for the understanding of light scattering from excitons. After a short description of the exciton states, the basic equations for the exciton dynamics are derived. This includes a discussion of the exciton–photon interaction and the concept of exciton–polaritons as a coupled system of exciton and photon in a solid. Concerning indirect excitons, it will be shown how their dynamics is determined in addition by the exciton–phonon interaction. The last part of this section is intended as a comprehensive overview on excitonic relaxation processes by phonons, since this is of relevance for the studies in this work. Here quantitative results for the scattering rates in three– and two–dimensional systems are given. To allow a unified treatment of all these aspects of exciton physics, the formalism of second quantization is used throughout in this chapter.

4.1 Excitons in Three Dimensions

4.1.1 Exciton States in Second Quantization

The starting point of our considerations is the simple two–band model of a semiconductor schematically shown in Fig. 4.1 [93]. The dependence of the single particle energies on the wavevector k is given by the dispersion relations of the conductance and valence band $E_c(k)$ and $E_v(k)$. In the following, they will be assumed to be of parabolic form

$$E_{c,v}(k) = \frac{\hbar^2 k^2}{2m_{e,h}} \,, \tag{4.1}$$

where m_e and m_h are the effective electron and hole masses. This simple relation is valid only for a few materials, like Cu_2O, while in most semiconductors the symmetry of the electron states requires more complicated expressions for the dispersion relations. The effective mass may be anisotropic as in germa-

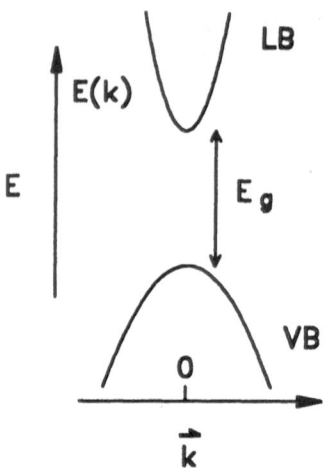

Fig. 4.1
Two–band model of a semiconductor showing schematically the dispersion of the single particle states of valence and conduction band.

E_g denotes the energy of the band edge.

nium or the silver halides (compare Sect. 6.2) or may show the so called 'warping' of the valence band in the III–V and II–VI compounds [94].

The Hamiltonian operator of the two–band model is given by [95, 93, 96]

$$H = H^{(1)} + H^{(2)} \tag{4.2}$$

$$H^{(1)} = \sum_{k,\sigma} E_c(\boldsymbol{k}) c^{\dagger}_{k,\sigma} c_{k,\sigma} - \sum_{k,\sigma} E_v(\boldsymbol{k}) d^{\dagger}_{k,\sigma} d_{k,\sigma} \tag{4.3}$$

$$- \sum_{k,k',q} \sum_{\sigma,\sigma'} V_q c^{\dagger}_{k,\sigma} c_{k-q,\sigma} d^{\dagger}_{k',\sigma'} d_{k'+q,\sigma'}$$

$$H^{(2)} = \frac{1}{2} \sum_{k,k',q} \sum_{\sigma,\sigma'} V_q \left\{ c^{\dagger}_{k,\sigma} c^{\dagger}_{k',\sigma'} c_{k'+q,\sigma'} c_{k-q,\sigma} d^{\dagger}_{k,\sigma} d^{\dagger}_{k',\sigma'} d_{k'+q,\sigma'} d_{k-q,\sigma} \right\} . \tag{4.4}$$

Here V_q denotes the Coulomb interaction

$$V_q = \frac{4\pi e^2}{\Omega q^2 \varepsilon_\infty} , \tag{4.5}$$

e the elementary charge, q the wavevetor, Ω the volume of the elementary cell and ε_∞ an effective dielectric constant describing the screening of the Coulomb field. $c^{\dagger}_{k,\sigma}$, $c_{k,\sigma}$ are the creation and annihilation operators of the conduction electrons, $d^{\dagger}_{k,\sigma}$ and $d_{k,\sigma}$ those of the hole states in the valence band. The index σ takes into account the spin degeneracy of the states.

The ground state of the two–band model is realized by an empty conduction and a fully occupied valence band with occupation numbers $\langle c^{\dagger}_{k,\sigma} c_{k,\sigma} \rangle = \langle d^{\dagger}_{k,\sigma} d_{k,\sigma} \rangle = 0$. Optical excitation transfers an electron from the valence band into the conduction band creating an electron–hole pair $c^{\dagger}_{k,\sigma} d^{\dagger}_{k',\sigma'}$. This two particle state, however, is not stable under the Coulomb interaction, but can be scattered into other states with the same total wavevector $k+k'$. Assuming that only a few electron–hole pairs are excited, the following superposition of electron–hole states, which is commonly called an *exciton*,

$$B^\dagger_{\alpha K \sigma \sigma'} = \sum_q F_{q\alpha}(K) c^\dagger_{q,\sigma} d^\dagger_{K-q,\sigma'} \tag{4.6}$$

represents approximately the true eigenstates of the crystal, provided the coefficients $F_{q\alpha}(K)$ are chosen properly.

Each exciton state is determined by an index α characterizing the degrees of freedom of the relative motion of the electron–hole pair, the wavevector K, representing the center of mass motion of the exciton and spin quantum numbers σ, σ' of electron and hole. As we have neglected the exchange interaction, which will be introduced below, the spins are not coupled together. Obviously, $B^\dagger$ creates an exciton while the adjunct operator B destroys it.

For the case of parabolic bands the energy eigenvalues of the exciton states are given by

$$E_\alpha(K) = E_\alpha + \frac{\hbar^2 K^2}{2M^*} + E_g \ . \tag{4.7}$$

Here E_g denotes the band edge, the first term represents the energy of the relative motion and the second term is the exciton kinetic energy, given by the wavevector K and the exciton translational effective mass $M^* = m_e + m_h$.

The $F_{q,\alpha}(k)$ are the expansion coefficients of the exciton states into electron–hole pair states. They describe a unitary transformation with orthogonality relations [96]

$$\sum_q F^*_{q,\alpha}(k) F_{q,\alpha'}(k) = \delta_{\alpha\alpha'} \tag{4.8}$$

and the inverse relations

$$c^\dagger_{q,\sigma} d^\dagger_{q',\sigma'} = \sum_\alpha F^*_{q\alpha}(q+q') B^\dagger_{\alpha q+q' \sigma \sigma'} \ . \tag{4.9}$$

Furthermore, the following relation holds

$$F_{q,\alpha}(k) = F_{q-m_e k/M^*,\alpha}(0) \ . \tag{4.10}$$

The $F_{q,\alpha}(0)$ are solutions of the exciton equation

$$\left(\frac{\hbar^2 q^2}{2M^*} + E_g - E_\alpha \right) F_{q,\alpha}(0) - \sum_{q'} V_{qq'} F_{q',\alpha}(0) = 0 \tag{4.11}$$

from which one obtains by Fourier transformation the usual wavefunction of the relative motion of the electron–hole pair

$$\frac{\sqrt{N\Omega_0}}{(2\pi)^3} \int d^3q \, F_{q,\alpha}(0) e^{iq\cdot r} = \Phi_\alpha(r) \ . \tag{4.12}$$

The exciton operators transform the two–band Hamiltonian into the simple form

$$H_{ex} = \sum_{K,\alpha,\sigma\sigma'} E_\alpha(K) B^\dagger_{\alpha K \sigma\sigma'} B_{\alpha K \sigma\sigma'} \tag{4.13}$$

which closely resembles that of a system of independent harmonic oscillators (see below).

Up to now it was implicitly assumed that the extrema of the valence and the conduction band occur at the same point of the Brillouin zone (BZ) forming a *direct exciton*. This need not be the case, as is realized in many important semiconductors like germanium and silicon [97]. In this book we discuss another example for these *indirect* excitons, the silver halides, where the minimum of the conduction band is at the centre of the BZ (Γ point) and the maximum of the valence band is at the L point of the BZ at $k_0 = k_L$ (see Chap. 6). The theoretical description of these indirect exciton states results from that of the direct states by referring all wavevectors to the band extrema [98].

4.1.2 Degenerate States

In order to provide a quantitative description of exciton states, the effect of the exchange interaction also has to be taken into account in the two–band model. Then the electron and hole spins couple together to give the total spin of the exciton. For the simple case of $s_{1/2}$ particles this results in a singlet ($S = 0$) and a triplet ($S = 1$) spin state. In the general case, we also have to sum over the spin states in Eq. (4.6) resulting in

$$B^\dagger_{\alpha K} = \sum_{q,m,n} F^{m,n}_{q\alpha}(K) c^\dagger_{q,m} d^\dagger_{K-q,n} . \tag{4.14}$$

Here m and n run over the spin states of electron and hole. Introducing again the envelope functions, these exciton states are the solutions of a generalized effective mass equation [99]

$$\sum_{m',n'} \left[E_{mn,m'n'}(-i\nabla, K) - \frac{e^2}{\varepsilon |r|} \delta_{nn'} \delta_{mm'} \right. \tag{4.15}$$

$$\left. + J_{mn,m'n'}(K) \delta_M \delta(r) \right] \Phi^\alpha_{m'n'}(r) = E_\alpha(K) \Phi^\alpha_{m,n}(r) ,$$

whereby $E_{mn,m'n'}$ is the sum of the energies of the valence and conduction band, and $J_{mn,m'n'}(K)$ is the exchange interaction, which is non–zero only for states with spin–singlet character ($M = 0$).

Equation (4.15) shows that for degenerate bands each exciton state is a well–defined combination of the different band states. To determine the mixing coefficients, it is in most cases not necessary to completely solve the exciton equation. Rather, these coefficients are determined by the symmetry of the system and are given as the coupling coefficients tabulated in Ref. [100]. For the decomposition of the product representation using these tables one should keep in mind that the hole states transform like the time–inverse states of the valence band.

4.2 Dynamics of Excitons

The main topic of this section is the derivation of the basic quantum mechanical equations that govern the temporal dynamics of an exciton system under the influence of a light field. A significant part of these considerations were put forward in the seventies by the group of *Haken* [95, 101], but at that time connections to experiments were still lacking. Also, important aspects of the dynamics of exciton–polaritons have been found only recently [102], so that a comprehensive discussion seems to be appropriate here. In addition, the description will be extended to indirect exciton states allowing insights into the dynamics of resonant Raman scattering processes.

4.2.1 Commutation Relations

In order that the Hamiltonian (4.13) describes a system of independent harmonic oscillators, as might be expected from its structure, one has to require that the exciton creation and annihilation operators fulfil the Bose commutation relations

$$\left[B_{\alpha K \sigma \eta}, B^\dagger_{\alpha' K' \sigma' \eta'} \right] = \delta_{K,K'} \delta_{\alpha \alpha'} \delta_{\sigma \sigma'} \delta_{\eta \eta'} . \tag{4.16}$$

To simplify the notation we will drop all spin indices in the following.

The commutation relations for the exciton operators can be derived from the Fermi anticommutation relations of the creation and annihilation operators of the conduction band states

$$c_k c^\dagger_{k'} + c^\dagger_{k'} c_k = \delta_{k,k'} \tag{4.17}$$

$$c_k c_{k'} + c_{k'} c_k = 0 \tag{4.18}$$

$$c^\dagger_k c^\dagger_{k'} + c^\dagger_{k'} c^\dagger_k = 0 \tag{4.19}$$

and from the equivalent relations for the hole operators. One obtains [95]

$$\left[B_K, B^\dagger_{K'} \right] = \delta_{K,K'} \tag{4.20}$$

$$- \sum_q F^*_q(K') F_q(K) d^\dagger_{K-q} d_{K'-q} - \sum_q F^*_{K'-K+q}(K') F_q(K) c^\dagger_q c_{K'-K+q}$$

$$\left[B_K, B_{K'} \right] = 0 \tag{4.21}$$

$$\left[B^\dagger_K, B^\dagger_{K'} \right] = 0 \tag{4.22}$$

The second term in Eq. (4.20), which contains the deviation from the Bose character of the exciton states, obviously depends on the density of electrons and holes. Therefore, only in the limit of weak excitation can this term be neglected and the excitons become bosons. At high densities and —most important for applications in nonlinear optical spectroscopy— for the description of higher order effects, this approximation breaks down. Here it was shown in recent years that a quite different description, the so called *semiconductor Bloch equations* allows a more accurate description of these effects [93,103–105]. In this book, however, we are interested in linear processes involving single excitons and the boson–like behaviour of excitons remains a valid concept allowing a substantial simplification of the exciton dynamics.

4.2.2 Exciton–Photon Interactions

The fundamental interaction of light with the two–band solid proceeds via the creation and annihilation of excitons. The most general form of this coupling is given by the $A \cdot p$ term of the Hamiltonian [106]. For most purposes it is sufficient to consider the dipole approximation of this interaction given by

$$H'_{\text{ex,phot}} = \sum_{q,\lambda,\alpha} i\epsilon_{q,\lambda} \cdot \left[(G_\alpha(q,\lambda)B_{q,\alpha} - G^*_\alpha(q,\lambda)B^\dagger_{-q,\alpha})(a_{-q,\lambda} + a^\dagger_{q,\lambda}) \right] . \quad (4.23)$$

Here G denotes the vectorial exciton–photon coupling coefficient

$$G(q) = E_\alpha \left(\frac{2\pi|\phi_\alpha(0)|^2}{\hbar c|q|} \right)^{1/2} M_{\text{cv}} , \quad (4.24)$$

where $c = c_{\text{vac}}/\sqrt{\varepsilon_\infty}$ is the velocity of light in the semiconductor considered as a dielectric medium with an effective dielectric constant ε_∞. M_{cv} represents the electronic band–to–band transition dipole moment per elementary cell of the crystalline solid and $\epsilon_{q,\lambda}$ are the polarization states of the light field.

For materials with dipole–forbidden transition one has to extend the expansion of the interaction operator to higher terms, e.g., to the quadrupolar coupling given by [107]

$$G^Q_\alpha(q,\lambda) = E_\alpha \left(\frac{2\pi|\phi_\alpha(0)|^2}{\hbar c|q|} \right)^{1/2} q \cdot M^Q_{\text{cv}} . \quad (4.25)$$

Here M^Q_{cv} denotes the quadrupolar transition tensor which is given by the expectation value of the dyadic product of the position and momentum operators.

In Eq. (4.23) the terms that are constructed from two annihilation or creation operators describe virtual processes, while those containing an annihilation and a creation operator describe real absorption processes. In the so called 'rotating–wave approximation' only the latter are of importance [93].

In the polariton problem care has to be taken as all four terms are relevant [108].

The temporal dynamics of the exciton–photon system is determined by the Heisenberg equation of motion for the exciton operators

$$\frac{d}{dt}B^\dagger(K) = \frac{i}{\hbar}\left[H, B_K^\dagger\right] . \tag{4.26}$$

In the Bose approximation one obtains

$$\frac{d}{dt}B^\dagger(K) = \frac{i}{\hbar}E(K)B_K^\dagger - \frac{1}{\hbar}\sum_{q,\lambda}G(q,\lambda)a_{q,\lambda}\delta_{K,q} . \tag{4.27}$$

Here we can deduce from the Kronecker–delta that only those exciton states with $K = q$ interact with light. As the wavevector of light is small compared to the extension of the BZ only excitons at $K \simeq 0$ interact directly with photons.

The equation of motion for the exciton operators has to be supplemented by the equation of motion for the photon field. This is derived from the Hamiltonian of the free field

$$H_{\text{phot}} = \sum_{q,\lambda}\hbar\omega_q a_{q\lambda}^\dagger a_{q\lambda} \tag{4.28}$$

and the interaction term (Eq. (4.23)) leading to

$$\frac{d}{dt}a_{q,\lambda}^\dagger = i\omega_q a_{q,\lambda}^\dagger + \frac{1}{\hbar}\sum_K G^*(K,\lambda)B_{K,\lambda}^\dagger\delta_{q,\lambda} . \tag{4.29}$$

As is well known from the work of *Haken* [95], the two relations (4.27) and (4.29) exactly correspond to the equation of motion of a system of independent harmonic oscillators. For linear processes at low excitation densities therefore the exciton dynamics can be calculated by well–known methods for treating harmonic oscillator systems [109, 110].

4.2.3 Direct Excitons: Polaritons

4.2.3.1 Polariton States. In the last section the exciton–photon interaction was treated as a small perturbation. For direct semiconductors, however, this assumption is not valid as the excitons also represent an electronic polarization wave which interacts strongly with the light field at the crossing of their dispersion relations. This interaction leads therefore to coupled states, which are called exciton–polaritons [102]. The use of this polariton concept allows a significant simplification of the dynamics of direct excitons.

The transformation from the uncoupled exciton and photon system to the coupled polariton states was first introduced by *Hopfield* [111]. A derivation based on a quantum mechanical microscopic picture was later given by *Bassani* and coworkers [102].

The polariton states are given by the following transformation law

$$A^\dagger_{K,\lambda,l} = \Pi^l_0(K,\lambda)a^\dagger_{K,\lambda} + \Phi^l_0(K,\lambda)a\dagger_{-K,\lambda} \qquad (4.30)$$
$$+ \sum_\alpha \Pi^l_\alpha(K,\lambda)B^\dagger_{K,\alpha} + \sum_\alpha \Phi^l_0(K,\lambda)B_{-K,\alpha}$$

whereby the summation over the exciton states α includes only those states with a non–zero component of the transition moment along the polarization vector of the photons ϵ_λ. It is therefore advantageous to choose a polarization basis for the light field which leads to a maximal decoupling of the exciton states.

The transformation coefficients are calculated from the requirement that the polariton operators fulfil Bose commutation relations and diagonalize the total Hamiltonian $H = H_{ex} + H_{phot} + H_{ex-phot}$. Here the $\mathbf{A}^2$ term of the exciton–photon interaction also has to be taken into account; it gives the following contribution to $H_{ex-phot}$ [102]

$$H''_{ex-phot} = \sum_{q,\lambda,\alpha} G''_\alpha(q,\lambda)(a^\dagger_q + a_{-q})(a^\dagger_{-q} + a_q) \qquad (4.31)$$

with

$$G''_\alpha(q,\lambda) = G^2_\alpha(q,\lambda)/E_\alpha . \qquad (4.32)$$

Note that here only the transverse eigenmodes of the exciton and photon field are considered [102].

Then the following transformation coefficients are obtained

$$\Pi^l_0(K,\lambda) = 1/2\,[cK + \Omega_l]\,\gamma(K,\Omega_l,\lambda)/\sqrt{cK\Omega_l} \qquad (4.33)$$
$$\Phi^l_0(K,\lambda) = 1/2\,[cK - \Omega_l]\,\gamma(K,\Omega_l,\lambda)/\sqrt{cK\Omega_l} \qquad (4.34)$$
$$\Pi^l_\alpha(K,\lambda) = i\epsilon_{K,\lambda}\cdot G^*_\alpha(K,\lambda)\sqrt{\frac{c_{vac}K}{\Omega_l}}\cdot\frac{\gamma(K,\Omega_l,\lambda)}{[\hbar\Omega_l - E_\alpha(K)]} \qquad (4.35)$$
$$\Phi^l_\alpha(K,\lambda) = -i\epsilon_{K,\lambda}\cdot G_\alpha(K,\lambda)\sqrt{\frac{c_{vac}K}{\Omega_l}}\cdot\frac{\gamma(K,\Omega_l,\lambda)}{[\hbar\Omega_l + E_\alpha(K)]} \qquad (4.36)$$

with

$$\gamma(K,\Omega,\lambda) = \frac{1}{\left[1 + \sum_\alpha \dfrac{4\varepsilon_\infty\hbar cKE_\alpha(K)|\epsilon_{K,\lambda}\cdot G_\alpha(K,\lambda)|^2}{(\hbar^2\Omega^2 - E_\alpha(K)^2)^2}\right]^{1/2}} . \qquad (4.37)$$

The polariton eigenfrequencies Ω_l are the solutions of the dispersion equation

$$\varepsilon(K,\Omega) = \frac{K^2 c^2_{vac}}{\Omega^2} = \varepsilon_\infty + \sum_\alpha \frac{4\varepsilon_\infty\hbar cK|\epsilon_{K,\lambda}\cdot G_\alpha(K,\lambda)|^2}{E_\alpha(K)\left(E_\alpha(K)^2 - \hbar^2\Omega^2\right)} . \qquad (4.38)$$

The relations show that the coupling between exciton and photon states is determined by the square of the coupling coefficient. Therefore the exciton oscillator strengths are introduced as convenient dimensionless parameters. In the general case they depend both on the direction of wavevector and the polarization of the photon. In the case of dipole coupling they are defined as [108]:

$$f_\alpha^D = \frac{\varepsilon_\infty}{\pi} \frac{\hbar c K}{E_\alpha^3(K)} |\epsilon_{K,\lambda} \cdot G_\alpha(K,\lambda)|^2 \qquad (4.39)$$

while for quadrupolar coupling the following relation holds

$$f_{K,\lambda}^Q = \frac{4\sqrt{\varepsilon_\infty}\hbar |\epsilon_{K,\lambda} \cdot G_\alpha(K,\lambda)|^2}{c_{vac} K E_\alpha(K)} . \qquad (4.40)$$

With these definitions, the dispersion equation of a quadrupole exciton polariton is [112]

$$\varepsilon(K,\Omega) = \frac{K^2 c_{vac}^2}{\Omega^2} = \varepsilon_\infty + \sum_\alpha \frac{f_{K,\lambda}^Q c_{vac}^2 K^2}{\left(E_\alpha(K)^2 - \hbar^2\Omega^2\right)} . \qquad (4.41)$$

Typical oscillator strengths for dipole transitions are of the order of 10^{-3}, while quadrupolar transitions have oscillator strengths of 10^{-9}.

4.2.3.2 Eigenfunctions of the Polariton Field.

In the polariton picture, the rate of scattering processes like Raman and Brillouin scattering by phonons is calculated by first–order perturbation theory applying *Fermi's golden rule* [113]. This requires the matrix elements of the corresponding interaction operators between the eigenfunctions of the polariton modes. These eigenfunctions can be obtained quite generally via the relation

$$\tilde{\Phi}(K,\lambda,l) = A_{K,\lambda,l}^\dagger \tilde{\Phi}_0 . \qquad (4.42)$$

Here $\tilde{\Phi}_0$ denotes the ground state of the polariton field. As was already pointed out by *Hopfield* (see also [108]) this ground state differs from the electromagnetic vacuum state (the ground state without coupling) because of the zero-point fluctuations of the exciton–polariton waves. The ground state can be obtained by the relation

$$A_{K,\lambda,l}\tilde{\Phi}_0 = 0 \qquad (4.43)$$

from the ground state Φ_0 without coupling. $\tilde{\Phi}_0$ differs from Φ_0 by the admixture of virtual exciton and photon pairs. In lowest order, when only one exciton and one photon needs to be considered and, furthermore, assuming that a single exciton state interacts with photons of defined polarization state ϵ_λ, one obtains [108]

$$\tilde{\Phi}_0 = \left[1 + \frac{1}{2}\sum_K \sum_{l,m=1}^{2} G_{lm}(K) C_{Kl}^\dagger C_{-Km}^\dagger\right] \Phi_0 \qquad (4.44)$$

where we have set $C_{K1} = a_{K,\lambda}$ and $C_{K2} = B_K$. The coefficients $G_{lm}(K)$ are given by the polariton transformation coefficients

$$G_{11}(K) = G_{22}(K) = -\frac{\hbar(\Omega_1 + \Omega_2) - E(K) - cK}{\hbar(\Omega_1 + \Omega_2) - E(K) + cK} , \qquad (4.45)$$

$$G_{12}(K) = G_{21}(K) = -i \left(\frac{\varepsilon_\infty \hbar cK}{\pi f_D E(K)}\right)^{1/2} G_{11}(K) . \qquad (4.46)$$

Actually Eq. (4.44) is the first order term of an exponential expression which was recently shown to represent a special form of a squeezed state of the photon field [114].

Using $\tilde{\Phi}_0$ the polariton wavefunction can be written as

$$A^\dagger_{K,\lambda,l}\tilde{\Phi}_0 = \left[\tilde{\Pi}^l_0(K,\lambda)a^\dagger_{K,\lambda} + + \sum_\alpha \tilde{\Pi}^l_\alpha(K,\lambda)B^\dagger_{K,\alpha}\right]\tilde{\Phi}_0 . \qquad (4.47)$$

Here the renormalized coupling coefficients

$$\tilde{\Pi}^l_0(K,\lambda) = \frac{\Pi^l_0(K,\lambda)}{|\Pi^l_0(K,\lambda)|^2 + |\sum_\alpha \Pi^l_\alpha(K,\lambda)|^2} \qquad (4.48)$$

and

$$\tilde{\Pi}^l_\alpha(K,\lambda) = \frac{\Pi^l_\alpha(K,\lambda)}{|\Pi^l_0(K,\lambda)|^2 + |\sum_\alpha \Pi^l_\alpha(K,\lambda)|^2} \qquad (4.49)$$

have been introduced. They describe the contribution of the uncoupled photon ($\tilde{\Pi}^l_0(K,\lambda)$) and the uncoupled exciton states ($\tilde{\Pi}^l_\alpha(K,\lambda)$) to the polariton mode.

This description now provides a rather different picture of the so-called ABC problem [108] as the transformation from the outer vacuum into the crystal can be taken as a simple change in the ground state of the system going from the electromagnetic vacuum into the squeezed coupled vacuum state. This picture will be applied in the description of time-resolved Raman scattering of exciton-polaritons in Cu_2O in Chap. 7.

For the 1S quadrupole exciton-polariton in this material the dispersion relations and the coupling factors are plotted Fig. 4.2. Due to the extremely small oscillator strength of $f = 3.6 \cdot 10^{-9}$ [115] the anticrossing of the dispersion curves at resonance is not visible (see however Fig. 7.3). In contrast, appreciable mixing between excitons and polariton occurs over a much larger range of energies.

4.2.4 Indirect Excitons

When the extrema of the valence or the conduction band occur at different points in the BZ, the dispersion of the indirect excitons made up from the

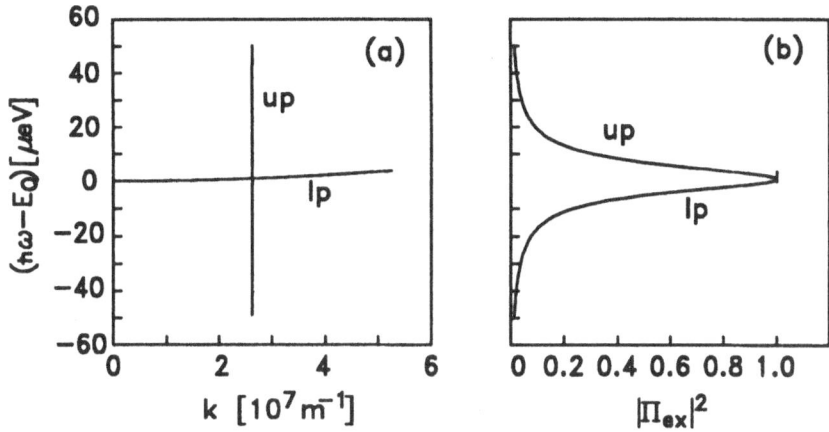

Fig. 4.2 Dispersion relations (a) and excitonic coupling factors (b) for exciton-polaritons in Cu_2O. E_Q: Exciton energy at $K = 0$; up, lp denote the upper and lower polariton branch; oscillator strength $f^Q = 3.6 \cdot 10^{-9}$.

band states has a minimum at $K \neq 0$. Due to the small photon wavevector, in an ideal crystal for these excitons no direct interaction with light is possible. Any optical interaction requires the participation of phonons in a higher order process thus supplying the necessary quasi–momentum, as is schematically depicted in Fig. 4.3. In the interaction Hamiltonian, besides the exciton-photon interaction $H_{ex-phon}$ with an electromagnetic field of frequency ω, the coupling of excitons with phonons of mode η, energy $\hbar\Omega^\eta$ and wavevector q has to be taken into account.

$$H_{ex-phot} = V^\eta(q)e^{-i(\Omega_q \cdot t)}b_q + h.c. \ . \tag{4.50}$$

The interaction strength given by $V^\eta(q)$ is determined by the specific electron-phonon coupling mechanism.

Time–dependent second order perturbation theory with the interaction operator

$$H_{int} = H_{ex-phot} + H_{ex-phon} \tag{4.51}$$

gives for the excitation probability of an exciton with wavevector K [99]

$$P_{\Psi_0 \to \Psi_K^{j,l}} = \frac{2\pi}{\hbar} \left| \sum_\lambda \frac{<\Psi_K^{j,l}|H_{int}|\lambda><\lambda|H_{int}|\Psi_0>}{E^{j,l}(K) - E(\Psi_0) - \hbar\omega} \right|^2 \delta(E_\lambda - E(\Psi_0) - \hbar\omega \pm \hbar\Omega^\eta) \ . \tag{4.52}$$

Here $E(\Psi_0)$ denotes the energy of the crystal ground state, E_λ the energy of the intermediate exciton states, $\hbar\omega$ the incident photon energy, and $\hbar\Omega^\eta$ the energy of the momentum conserving phonon (upper sign: phonon emission, lower sign: phonon absorption). The index j denotes degenerate states at the same k–vector in the star of the band minimum, the different k-vectors numbered by the index l.

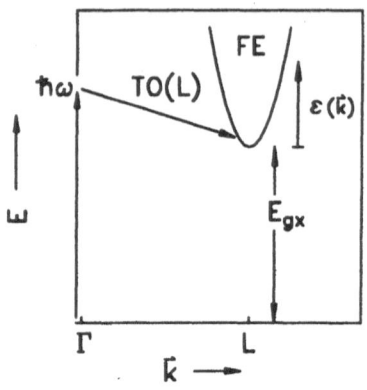

Fig. 4.3
Schematic dispersion of an indirect exciton for AgBr, in which the minimum of the exciton dispersion is at the L point of the BZ. $\hbar\omega$: photon energy; $\epsilon(\boldsymbol{k})$: exciton kinetic energy; E_{gx}: exciton band edge.

Inserting the wavefunction (Eq. (4.14)) results in

$$P_{\Psi_0 \to \Psi_K^{j,l}} = \frac{2\pi}{\hbar}\left(\frac{eA_0}{mc}\right)^2 \left|\sum_{k,m,n} F_{k,j,l}^{m,n}(\boldsymbol{K})\epsilon_\lambda \cdot \boldsymbol{M}_{m,n}^{\eta}(\boldsymbol{k},\boldsymbol{K})\right|^2 \quad (4.53)$$

$$\times \delta(E^{j,l}(\boldsymbol{K}) - E(\Psi_0) - \hbar\omega \pm \hbar\Omega^\eta)$$

with A_0 the amplitude of the vector potential of the electromagnetic wave. $\boldsymbol{M}_{m,n}^{\eta}(\boldsymbol{k},\boldsymbol{K})$ is the effective transition moment of the indirect band–to–band transition given by

$$\epsilon_\lambda \cdot \boldsymbol{M}_{m,n}^{\eta}(\boldsymbol{k},\boldsymbol{K}) =$$

$$\sum_\lambda \frac{\langle c,m,\boldsymbol{k},n^\eta|\epsilon_\lambda \cdot \boldsymbol{p}|\lambda,\boldsymbol{k},n^\eta\rangle\langle\lambda,\Gamma,n^\eta|V^\eta(\pm(\boldsymbol{k}-\boldsymbol{K}))|v,n,\boldsymbol{k}-\boldsymbol{K},n^\eta\mp 1\rangle}{E_v(\boldsymbol{k}-\boldsymbol{K}) - E_\lambda(\boldsymbol{k})\mp\hbar\Omega^\eta}$$

$$+\frac{\langle c,m,\boldsymbol{k},n^\eta\pm 1|V^\eta(\pm\boldsymbol{k}-\boldsymbol{K}))|\lambda,\boldsymbol{k}-\boldsymbol{K},n^\eta\rangle\langle\lambda,\boldsymbol{k}-\boldsymbol{K},n^\eta|\epsilon_\lambda \cdot \boldsymbol{p}|v,n,\boldsymbol{k}-\boldsymbol{K},n^\eta\rangle}{E_c(\boldsymbol{k}) - E_\lambda(\boldsymbol{k}-\boldsymbol{K})\pm\hbar\Omega^\eta} .$$

$$(4.54)$$

In this expression, each term consists of two virtual processes, the direct transition by the electron–photon interaction operator $\epsilon_\lambda \cdot \boldsymbol{p}$ ($\boldsymbol{p}$ momentum operator, e polarization vector of the electromagnetic field, momentum selection rule $\Delta k = 0$) and a phonon scattering process with the electron–phonon operator V^η of the phonon mode η with occupation number n^η and energy $\hbar\omega^\eta$ providing the momentum transfer $(\pm\boldsymbol{k} - \boldsymbol{K}))$ (upper sign: phonon emission, lower sign: phonon absorption). The first term represents valence band (in the 'missing electron' scheme), the second conduction band scattering, both together adding up coherently to the total rate.

For allowed transitions the effective transition moment is independent of the wavevector of the final exciton state $\boldsymbol{k}$. This requires both the direct dipole transition and the phonon scattering to be allowed. Forbidden indirect transitions which are usually very weak are not considered here (see e.g. [116]).

In this case the absorption coefficient of an exciton state j via momentum conserving phonons of mode η is given by

$$\alpha_{j,l}^{\eta}(E,\epsilon_\lambda) = \frac{\hbar}{E} \cdot \frac{2e^2}{nc_L m_e^2} \cdot \left(\frac{2M_{ex}^*}{\hbar^2}\right)^{3/2} \cdot \left|\sum_{m,n} \Phi_{m,n}^{j,l}(0)\epsilon_\lambda \cdot M_{m,n}^{\eta}\right|^2$$

$$\times \left(n^{\eta} + \frac{1}{2} \pm \frac{1}{2}\right) \cdot \sqrt{E - E_{gx}^i \mp \hbar\Omega^{\eta}}$$

(4.55)

whereby the index l discriminates the exciton states in different valleys. $E = \hbar\omega$ is the incident photon energy, n the refractive index, c_L the velocity of light, and m_e the free electron mass. $|\Phi(0)|^2$ denotes the square of the exciton envelope function at $r = 0$ which is for a $1S$ state given by $|F(0)|^2 = 1/\pi a_{ex}^3$ with a_{ex} the exciton Bohr radius. n_{phon}^{η} is the thermal phonon occupation, given by a Bose–Einstein distribution

$$n_{phon}^{\eta}(\hbar\omega_q^{\eta}, T) = \frac{1}{\exp[\hbar\omega_q^{\eta}/kT] - 1} \cdot$$

(4.56)

To treat the dynamics of an indirect exciton system systematically, one has to include the dynamics of the momentum conserving phonons, too. In the limit of vanishing excitation intensity only spontaneous phonon processes are important. Since at low intensities also the virtually excited exciton states do not change the optical response of the material, the phonon dynamics can be neglected [117, 118]. In this approximation the interaction processes can be described by an effective transition matrix element

$$G_{eff}^{j,l}(q,\eta) = \hbar\omega_q \left(\frac{2\pi \left|\sum_{m,n} \Phi_{m,n}^{j,l}(0)\epsilon_{q,\lambda} \cdot M_{m,n}^{\eta}\right|^2 \left(n_{phon}^{\eta} + \frac{1}{2} \pm \frac{1}{2}\right)}{\hbar\omega_q}\right)^{1/2}$$

(4.57)

where $M_{m,n}^{\eta}$ is given by Eq. (4.54). Most important is the fact that in this limit the phonons have no influence on the exciton dynamics, especially on the coherence properties of the exciton states. This has important consequences for the light scattering processes (see Sect. 5.3) and can be checked experimentally.

4.3 Excitons in Quantum Well Structures

4.3.1 Band Structure of III–V Compounds

In recent years, the properties of exciton states have attracted increased interest due to the possibility of reducing the dimensionality of a material system and to tailor two– one– and even zero–dimensional structures (see e.g. [119]). Here the main topic will be investigations of two–dimensional systems, which are commonly named 'quantum well' (QW) structures. These artificial crystals are made by growing successive layers of at least two different materials by means of special epitaxy methods like *molecular beam epitaxy* (MBE) or *metal organic vapour phase epitaxy* (MOVPE) (for a recent review see e.g. [120]). The two materials composing the alternating layers usually differ in their band edge energies. The first and most investigated material system is formed by the III–V compounds GaAs and mixed crystals from GaAs and AlAs. They have the unique property that the lattice constant is almost independent on the Al content thus allowing one to produce almost perfect samples. The growing direction of the two–dimensional material stack, which in most cases is a (001) crystallographic direction, will be assumed as the z axis of a cartesian coordinate system.

To calculate the single particle energies in such a layered structure in principle one has to perform a band structure calculation with a variable composition of the material in the z direction. This formidable task has not been performed up to now. Instead, the electronic states are calculated in the 'envelope function' approximation [121]. Here one assumes that the variation of the material system as a function of z can be modeled in a variation of the corresponding band edges acting like an external potential that quantizes the electronic states in the z direction. In the plane perpendicular to the growth direction, the crystal lattice is still periodic so that the Bloch states are retained. The most simple case for such a structure is the single quantum well shown schematically in Fig. 4.4 where the energy shifts are determined mainly by the effective masses of the electrons and holes. Due to the different masses of the *light hole* (lh) and *heavy hole* (hh) states the degeneracy of the bulk material at the Γ point is lifted, consistent with the symmetry reduction of the quantum well structure.

While the z component of the wavevector k loses all meaning, the states can still be characterized by the in–plane component of k, resulting in two–dimensional dispersion relations. For the conduction band the dependence of the energy on $k_\parallel$ is still parabolic in a first approximation. For the valence bands the resulting dispersion relations are quite complex and known only in a few cases theoretically [121, 122]. From these calculations it follows that the transverse hole mass for heavy holes is smaller than for light holes. This leads then to a crossing of the bands at some finite value of $k_\parallel$ resulting in a strong non–parabolicity of the valence band dispersion relations. By fitting the real dispersion to parabolic approximations near $k_\parallel = 0$, approximate effective masses can be obtained which depend strongly on the width of the quantum well L_z. For the heavy hole states this results in $m_{hh} \simeq 0.165$ for $L_z > 5$ nm and $m_{hh} \simeq 0.118$ for $L_z < 5$ nm, while for the light hole negative effective masses always result. This means that the maximum of the light hole dispersion occurs outside of the centre of the two–dimensional BZ at a wavevector of

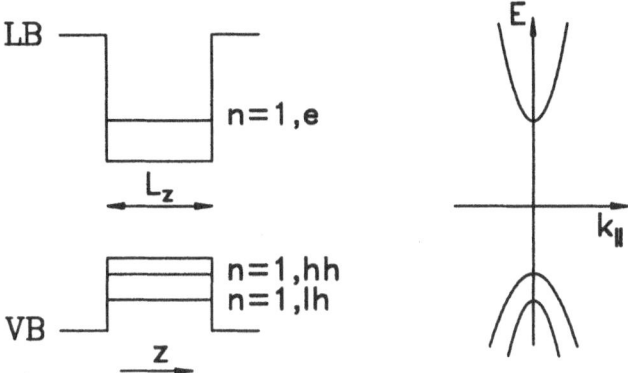

Fig. 4.4 Quantization of electron and hole energies in an one–dimensional quantum well according to effective mass theory. The right side shows the two–dimensional dispersion relations in the QW layer.

$k_\parallel \simeq 1.5 \cdot 10^6$ cm^{-1}. At the same wavevector the dispersion relations of heavy and light hole cross. Unfortunately there is so far no experimental evidence for this indirect transition.

4.3.2 Exciton States

In two–dimensional structures the Coulomb interaction is also able to lead to bound electron–hole pair states, the two–dimensional excitons. In an exactly two–dimensional system they correspond to the two–dimensional hydrogen problem [93], resulting in a fourfold binding energy and half the Bohr radius a_B^{2d} compared to a three–dimensional system.

In quantum well structures the quantized motion of electrons and holes along the z axis has to be taken into account. For this quasi–two-dimensional system only approximate solutions of the exciton problem exist. Most commonly a variational procedure is used to obtain binding energies and exciton wavefunctions (see e.g. [121]). A simple trial function for the exciton state is obtained by assuming a hydrogen–like relative motion and infinite potential barriers for the electron and hole. This leads to the expression

$$\Phi_{1s}(\rho, z_e, z_h) = N_{1s} \cdot \exp\left(-\frac{|\rho|}{a_B^{2d}}\right) \cos\left(\pi z_e / L_z\right) \cos\left(\pi z_h / L_z\right) \qquad (4.58)$$

where the effective Bohr radius a_B^{2d} is treated as a variational parameter.

N_{1s} is a normalization constant given by

$$N_{1s} = \sqrt{\frac{8}{\pi}} \cdot \frac{\Omega_0}{L L_z a_B^{2d}} \qquad (4.59)$$

whereby L denotes the length of the (square) quantum well layer. This wavefunction allows a rather accurate description of the excitonic state in quantum wells for thicknesses L_z which are smaller than the Bohr radius $a_B^{2d} \simeq 10$ nm [121] and will be used below to calculate exciton–phonon scattering rates in quantum wells.

Unfortunately, for the excitonic dispersion relation $E(K_{\parallel})$, which is most important to describe the exciton relaxation processes, no theoretical calculation has yet been performed. Therefore only qualitative models assuming a parabolic dispersion with the effective mass as unknown fitting parameter are available. The following considerations, however, show that this simple model might not be too bad for the description of the excitonic states. The exciton states are always made up from band states corresponding to a range of wavevectors, whereby k–values up to a maximum wavevector of the order of $1/a_B$ contribute. Assuming $a_B^{2d} = 10$ nm, which is quite reasonable for GaAs quantum wells, the range of wavevectors just covers the region of strong nonparabolicity, averaging over all details of the valence band dispersion which therefore can be approximated by an average value. This average value for the exciton translational mass, however, can be different from the sum of electron and hole masses, as is well known for the bulk of most III-V and II-VI semiconductors from theoretical calculations of the exciton dispersion [123] and from experimental investigations by resonant Brillouin scattering [113]. In going from the T_d bulk symmetry to the D_{2d} symmetry of the quatum well by introducing an external potential, one can expect that the exciton masses of the hh and lh states change only in higher order. Indeed, exactly this behaviour was recently found in biaxially strained epitaxial layers of ZnSe on GaAs substrate, showing the same symmmetry reduction. In the case of GaAs QW structures, applying this picture would result in a hh translational exciton mass of $M_{hh}^{*} \simeq 1.0 m_0$ [123], much larger than the sum of the carrier masses $m_e + m_{hh} = 0.234 m_0$ (see Sect. 4.3.1).

4.4 Relaxation Processes of Excitons

At low exciton densities all excitonic relaxation processes in a pure crystal originate in the interaction of the electronic states with lattice vibrations. Any investigation of the dynamics of excitons is therefore incomplete without a detailed understanding of the excitonic relaxation processes mediated by phonons. It is the purpose of this section to provide a comprehensive overview of the relaxation mechanisms by interaction with phonons as these are most important for the studies reported here. The foundations of the theoretical work were already laid by the papers of *Toyozawa* [98] and *Ansel'm* and *Firsov* [124] in the late fifties. Direct experimental investigations of these relaxation processes have been performed only recently, a great many by the method of light scattering. (for a review see e.g. [36]).

4.4.1 Exciton–Phonon Interactions

The exciton–phonon interaction Hamiltonian is given in lowest order by [26]:

$$H_{\text{ex,phon}} = \tag{4.60}$$
$$\sum_{K,\alpha}\sum_{K',\alpha'}\sum_{q,\lambda} i Q_{\sigma,q}(K,\alpha,K',\alpha') B^\dagger_{K,\alpha} B_{K',\alpha'} (b_{q,\sigma} - b^\dagger_{-q,\sigma}) \delta(q - K + K')$$

where we have also used the formalism of second quantization. The operator $b^\dagger_{q,\sigma}$ creates a phonon with wavevector q and mode σ, $Q_{\sigma,q}(K,\alpha,K',\alpha')$ is the exciton–phonon coupling coefficient that depends of course on the detailed interaction mechanism. The δ–function takes into account the conservation of quasi–momentum during the scattering process.

4.4.1.1 Types of Exciton–Phonon Interactions. There are different types of exciton–phonon interactions differing in the coupling coefficients. For Wannier excitons, which are the only ones of interest here, the exciton–phonon interaction is composed of the electron– and hole–phonon interactions. The coupling coefficients are therefore of a general form and given by

$$Q_{\sigma,K-K'}(K,\alpha,K',\alpha') = \Xi^\sigma_c(K-K')q(\alpha_h(K-K'),\alpha,\alpha') \tag{4.61}$$
$$-\Xi^\sigma_v(K-K')q(\alpha_e(K-K'),\alpha,\alpha') \ .$$

Here $\alpha_e = m_e/M$ and $\alpha_h = m_h/M$ denote the relative electron and hole masses. The quantity $q(X,\alpha,\alpha')$ that is common to the different parts of the expression was introduced by *Ansel'm* and *Firsov* [124] and is a measure of the overlap of the wavefunctions of electron and hole in the exciton state. It depends on the wavevector of the scattered phonon and can be calculated as the Fourier transform of the exciton wavefunction. For the lowest and most important $1S$ state it is given by [98]

$$q(X,1S,1S) = \frac{1}{\left[1 + (|X|/2a_B)^2\right]^2} \tag{4.62}$$

with the exciton Bohr radius a_B . This relation shows that for exciton scattering by phonons only a limited range of phonon wavevectors between 0 and $\alpha_h/2a_B$ for electron, and $\alpha_e/2a_B$ for hole scattering is of importance.

The coupling coefficients $\Xi^\sigma_{c,v}(K-K')$ depend on the interaction mechanism and have the following form for the most important scattering mechanisms:

1. Deformation potential interaction with acoustic phonons (DP–scattering). The origin of this interaction is the shift of the band edges under an elastic strain induced by the lattice vibrations. The DP mechanism occurs in cubic crystals in the case of non-degenerate bands only for longitudinal phonons

$$\Xi^{D,LA}_{c,v}(K - K') = \sqrt{\frac{\hbar}{2\rho N \Omega_0 u_{LA}}} |K - K'|^{1/2} D_{c,v} . \qquad (4.63)$$

Here $D_{v,c}$ are the deformation potentials of valence and conduction band, i.e. the shift of the band edge due to unit dilatation of the lattice. u_{LA} denotes the longitudinal sound velocity, N is the number and Ω_0 the size of an elementary cell. ρ is the density of the material.

For degenerate bands, like the uppermost valence bands in semiconductors with the zinc blende structure with Γ_8 symmetry, the deformation potential interaction is more complicated and scattering of holes with both LA and TA modes occurs (see e.g. [94]).

2. Optical deformation potential in crystals with two or more atoms in the basis (OD–scattering). According to [94] this occurs only for band states at those points of the BZ where the optical phonons transform under the symmetry operations of the crystal like the unit representation, as e.g. is the case for the L–point in the sodium–chloride structure. It is given by the optical deformation tensor, or in the special case of a threefold rotation symmetry by the deformation vector $D_{od\sigma}$

$$\Xi^{od,\sigma}_{c,v}(K - K') = \sqrt{\frac{\hbar}{2\rho N \Omega_0 \omega_\sigma}} D_{od,\sigma} \cdot e_{K-K',\sigma} \qquad (4.64)$$

with $e_{K-K',\sigma}$ the shift vector of the phonon mode.

3. Fröhlich coupling with polar optical phonons in crystals with ionic character (polar optical (PO) scattering). Here the interaction matrix element is given by [26]

$$\Xi^{po,LO}_{c,v}(K - K') = \sqrt{\frac{2\pi \hbar \omega_{LO} e^2}{N \Omega_0}} \sqrt{\left[\frac{1}{\varepsilon_\infty} - \frac{1}{\varepsilon_0} \right]} \cdot |K - K'|^{-1} \qquad (4.65)$$

with $\hbar \omega_{LO}$ the LO–phonon energy, e the elementary charge, ε_0 and ε_∞ the relative dielectric constants at zero and infinite frequencies.

4. Piezoelectric scattering by acoustic phonons (PE–scattering). This type of interaction occurs in crystals without inversion symmetry both for longitudinal and transversal modes ($\sigma =$LA,TA)

$$\Xi^{pe,\sigma}_{c,v}(K - K') = \frac{2 e e_{14}}{\varepsilon_0 \varepsilon_0} \sqrt{\frac{\hbar}{2\rho N \Omega_0 u_L}} \frac{1}{|K - K'|^{1/2}} A_\sigma(K - K') . \qquad (4.66)$$

e_{14} is the piezo–elastic coefficient and ε_0 the dielectric constant of the vacuum. The quantity $A_\sigma(K - K')$ takes into account the dependence of the piezo–electric coupling on the direction of the scattering phonon $Q = K - K'$ [125].

For semiconductors crystallizing in the zinc–blende structure Q is referred to the [001]–direction of the crystal. Then A can be expressed by the polar angles Θ and Φ [125]:

$$A(\Theta, \Phi)_{\text{LA}} = \frac{3}{2} \sin^2 \Theta \cos \Theta \sin 2\Phi \tag{4.67}$$

$$A(\Theta, \Phi)_{\text{TA}}^1 = \frac{1}{2}(3 \sin^2 \Theta - 2) \sin \Theta \sin 2\Phi \tag{4.68}$$

$$A(\Theta, \Phi)_{\text{TA}}^2 = \frac{1}{2} \sin^2 \Theta \cos 2\Phi \tag{4.69}$$

for the LA and the two TA modes.

4.4.2 Energy Relaxation Times of Excitons

The interactions of excitons with phonons lead to the possibility of transitions between different exciton states by emission and absorption of phonons. As these scattering processes are always connected with a loss of energy they represent a population relaxation mechanism for the exciton states. The scattering rate from an exciton state K, α can be calculated by 'Fermi's golden rule' (see e .g. [106]):

$$\Gamma_{\text{phon}}(K, \alpha) = \frac{2\pi}{\hbar} \sum_{K', \alpha'} \sum_{q, \lambda} |Q_{\sigma, q}(K, \alpha, K', \alpha')|^2 \tag{4.70}$$

$$\times \left[n_{\text{phon}, \sigma}(\hbar\omega_q) \delta(E_\alpha(K) - E_{\alpha'}(K') + \hbar\Omega_q) \delta(K - K' + q) \right.$$
$$\left. + (n_{\text{phon}, \sigma}(\hbar\omega_q) + 1) \delta(E_\alpha(K) - E_{\alpha'}(K') - \hbar\Omega_q) \delta(K - K' - q) \right] .$$

This expression was obtained by averaging over a distribution of phonons with occupation numbers $n_{\text{phon}, \sigma}(\hbar\omega_q)$. In thermal equilibrium it is given by a Bose–Einstein distribution

$$n_{\text{phon}}(\hbar\omega_q, T) = \frac{1}{\exp[\hbar\omega_q / kT] - 1} . \tag{4.71}$$

For energy–dispersive phonons, such as acoustic modes, whose frequency depends on the wavevector, besides the total rate, the spectral scattering rate can be calculated which determines the transitions between excitons of well–defined energy. Here the summation in Eq. (4.70) has to be restricted to final states of the same energy.

The assumptions that have to be made to derive the 'golden rule' are equivalent to the Markov approximation [9]. This implies that Eq. (4.70) does not describe the excitonic relaxation processes properly if either the energy uncertainty due to the relaxation rate approaches the order of the

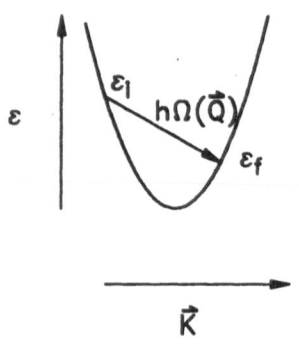

Fig. 4.5
Schematic picture of the intra–band relaxation of excitons by phonons of energy $\hbar\Omega$ and wavevector Q. ϵ_i, ϵ_f: exciton kinetic energies of initial and final state.

phonon energies themselves, or the time resolution with which the relaxation processes are observed experimentally is of the order of the reciprocal phonon frequency, since it is well known that in this regime the Markov approximation is not valid [118, 126]. In the systems which are of interest here, however, the discussion of the experiments will show that the 'golden rule' can still be applied.

In the following, some quantitative results for the scattering rates will be given. In all cases we will assume a simple parabolic exciton dispersion, as shown schematically in Fig. 4.5, which is characterized by an effective exciton mass M^* and where the scattering occurs only inside this band.

4.4.2.1 Deformation Potential Scattering. In this case we consider the intraband scattering with long–wavelength acoustic phonons via the deformation potential coupling. For the spectral scattering rate $\Gamma(\epsilon, \epsilon')$ for a transition between an exciton state with kinetic energy ϵ to all states at energy ϵ' the following expression results

$$\Gamma^{S,AS}_{ac,LA}(\epsilon, \epsilon') = \frac{3}{4} c_{ac} \frac{(\epsilon - \epsilon')^2}{\sqrt{\epsilon/\epsilon_0^*}} \left(n_{phon}(|\epsilon - \epsilon'|, T) + \frac{1}{2} \pm \frac{1}{2} \right) \qquad (4.72)$$

with the upper sign applying to Stokes $(S, \epsilon' < \epsilon)$ and the lower sign to anti–Stokes scattering $(AS, \epsilon' > \epsilon)$. Here $\epsilon_0^* = M^* u_{LA}^2/2$ with the sound velocity u_{LA}. The final state energy ϵ' varies for Stokes processes in the range

$$\min\left(\epsilon, \epsilon - 4\epsilon_0^* \left(\sqrt{\epsilon/\epsilon_0^*} - 1 \right) \right) \le \epsilon' \le \epsilon \qquad (4.73)$$

while for anti–Stokes scattering the range of final energies is given by

$$\max\left(\epsilon, \epsilon - 4\epsilon_0^* \left(1 - \sqrt{\epsilon/\epsilon_0^*} \right) \right) \le \epsilon' \le \epsilon + 4\epsilon_0^* \left(1 + \sqrt{\epsilon/\epsilon_0} \right) . \qquad (4.74)$$

In these equations $|\epsilon - \epsilon'|$ represents the phonon energy and T is the lattice temperature.

Neglecting the overlap–factors q, the quantity c_{ac} is constant and given by

$$c_{ac} = \frac{1}{4\pi} \frac{(D_c - D_v)^2}{\hbar^4 \rho u_{LA}^5} . \tag{4.75}$$

Otherwise the difference in the deformation potentials has to be replaced by $q(\alpha_h Q)D_c - q(\alpha_e Q)D_v$ with $Q = |\epsilon - \epsilon'|/\hbar u_{LA}$, the wavevector of the scattered phonon.

4.4.2.2 Piezoelectric Scattering.

For this mechanism results similar to the DP scattering are obtained. Here the spectral scattering rate is

$$\Gamma_{pe,\sigma}^{S,AS}(\epsilon, \epsilon') = \frac{3}{4} c_{pe,\sigma} \frac{1}{\sqrt{\epsilon/\epsilon_{0,\sigma}^*}} \left(n_{phon}(|\epsilon - \epsilon'|, T) + \frac{1}{2} \pm \frac{1}{2} \right) \tag{4.76}$$

whereby scattering by longitudinal ($\sigma = LA$) or transversal ($\sigma = TA$) phonons is possible. The range of possible final state energies is the same as for the DP mechanism, only the sound velocity is that of the corresponding phonon mode. The constant c_{pe} is given by

$$c_{pe,\sigma} = \frac{1}{4\pi} \cdot \left(\frac{ee_{14}}{\epsilon_0 \varepsilon_0} \right)^2 \frac{1}{\rho \hbar^2 u_\sigma^3} \tag{4.77}$$

and the unit of energy $\epsilon_{0,\sigma}^* = M^* u_\sigma^2/2$.

4.4.2.3 Fröhlich Interaction.

The Fröhlich mechanism is responsible for intraband–scattering by long–wavelength polar optical phonons. Here the spectral scattering rate is given by

$$\Gamma_{LO}^{S,AS}(\epsilon, \epsilon') = c_{LO} \sqrt{\epsilon'}(\epsilon + \epsilon')\delta(\epsilon - \epsilon' - \hbar\Omega_{LO}) \left(n_{phon}(\hbar\Omega_{LO}, T) + \frac{1}{2} \pm \frac{1}{2} \right), \tag{4.78}$$

where energy conservation is taken into account by the δ–function. $\hbar\Omega_{LO}$ is the dispersionless LO phonon energy. The constant c_{LO} is given by

$$c_{LO} = \left(\frac{1}{\varepsilon_\infty} - \frac{1}{\varepsilon_0} \right) \hbar\Omega_{LO}(m_h^* - m_e^*)^2 \frac{e^2 a_B^4}{4\pi\epsilon_0 \hbar^6} \sqrt{2M^*} \tag{4.79}$$

depending on the difference in the relative dielectric constants at infinite (ε_∞) and zero frequency (ε_0), the effective electron (m_e^*) and hole (m_h^*) masses and on the exciton Bohr radius a_B.

4.4.2.4 Intervalley Scattering.

In the case of indirect excitons, in addition to the scattering processes discussed up to now, there is the possibility of scattering between inequivalent valleys, the intervalley scattering processes. In this work we have to consider this relaxation mechanism in the case of the indirect L–exciton in silver bromide (see [42]). The energy dependence of the scattering rate is also given by a relation of the form (4.78), whereby the proportionality constant (here c_{iv}) is determined by the intervalley deformation potentials (for details see [22]).

4.4.2.5 The Total Scattering Rate. The total scattering rate that determines the exciton lifetime is obtained from the spectral rates by summation over all possible final exciton states. For the intervalley and polar optical scattering this results in expression (4.72) without the δ–function, for the acoustic phonon scattering the integration leads to

$$\Gamma_{ac}(\epsilon, T) = \int_{\epsilon'_{min}}^{\epsilon'_{max}} d\epsilon' \Gamma^S(\epsilon, \epsilon') + \int_{\epsilon'_{min}}^{\epsilon'_{max}} d\epsilon' \Gamma^{AS}(\epsilon, \epsilon') . \tag{4.80}$$

4.4.3 Exciton–Polariton Scattering

Thus far, only the scattering processes between exciton states have been discussed. But for direct excitons the polariton states, i.e. the mixed exciton and photon states are the relevant elementary excitations that undergo the relaxation processes. The scattering rates of polaritons are different from the exciton rates for two reasons:

1. The different dispersion relation $E(K)$ leads to a change in the phase space of the final states.
2. The admixture of photons to the exciton wavefunctions results in a reduction of the scattering matrix element, as only the excitonic part is relevant for the interaction with phonons.

The rate for phonon–polariton scattering can also be calculated using 'Fermi's golden rule', whereby only the coupling factor has to be transformed into polariton states using the transformation coefficients $\tilde{\Pi}^l$. This leads to

$$Q^{pol}_{\sigma,q}(K, l, K', l') = \sum_{\alpha, \alpha'} \tilde{\Pi}^l_\alpha(K) \tilde{\Pi}^{l'*}_{\alpha'}(K') Q_{\sigma,q}(K, \alpha, K', \alpha') . \tag{4.81}$$

This relation shows that the polariton scattering rate is determined by the exciton admixture in both the initial and final state.

4.4.4 Relaxation Processes of Two–Dimensional Excitons

As in three–dimensional systems, the dynamical behaviour of two–dimensional excitons is governed by the phonon relaxation processes. At low temperatures and exciton energies small compared to the optical phonon energy, intraband relaxation by acoustic phonons dominates. Relaxation rates for excitons in QW have been calculated earlier by various groups [127, 128], but these calculations do not give quantitative results that can be compared to experiment. In particular, the spectrally discriminated scattering rates that determine the exciton energy distribution were not determined. Furthermore, in these calculations the deformation potential scattering of the holes in GaAs was treated assuming an isotropic deformation potential, not taking into account the symmetry of the hole states. As will be shown below, inclusion of these effects will change the scattering rates appreciably.

The calculation summarized here follows closely the general procedure given by *Takagahara* [127], and only the main results will be presented.

The starting point is the expression for the relaxation rate (Eq. (4.70)) whereby wavevector conservation is only required for the component of the scattered phonon wavevector Q in the QW plane ($Q_\parallel$), while the component Q_z perpendicular to the layer is not restricted.

To calculate the matrix elements of the exciton–phonon interactions for the various scattering mechanisms $Q^{2d}_{\sigma,Q}(K_\parallel + Q_\parallel, K_\parallel)$ (σ: LA–DP, LA–PE, TA–PE) describing the scattering of an exciton with wavevector $K_\parallel + Q_\parallel$ into a final state with wavevector $K_\parallel$ by a phonon with wavevector Q from the corresponding bulk values, the electron and hole wavefunctions and the exciton envelope function have to be known. For the calculation we assume these functions of the form of Eq. (4.58); this greatly simplifies the calculations but still gives a good description of the QW exciton states. We further assume that the acoustic phonons are not quantized in the two–dimensional structure, i.e. the sound velocities of the well and barrier material are the same.

After a somewhat tedious calculation the following result for the exciton–phonon coupling factor is obtained

$$Q^{2d}_{\sigma,Q}(K_\parallel + Q_\parallel, K_\parallel) = \Xi_c(Q_\parallel, Q_z) \cdot q^{2d}(\alpha_h \cdot Q_\parallel, z) \qquad (4.82)$$
$$- \Xi_v(Q_\parallel, Q_z) \cdot q^{2d}(\alpha_e \cdot Q_\parallel, z) .$$

Here q^{2d} denotes an effective two–dimensional overlap–function

$$q^{2d}(X, Q_z) = \frac{\sin(Q_z L_z/2)}{Q_z L_z/2} \cdot \frac{1}{1 - (Q_z L_z/2\pi)^2} \cdot \frac{1}{\left[1 + \left(|X| a_B^{2d}/2\right)^2\right]^{3/2}} . \qquad (4.83)$$

The first two factors replace the strict wavevector conservation of three–dimensional systems, and the last factor is due to the finite overlap of electron and hole wavefunction.

For the conduction band contribution, the quantity Ξ_c is identical to that in the corresponding three–dimensional crystal and given by Eqs. (4.63) and (4.64). For the holes, we have to consider the deformation potential interaction in more detail. In the bulk crystal, it is given by the well-known *Bir-Pikus* Hamiltonian [94]

$$H_\epsilon = a \, \mathrm{tr} \, \underline{\epsilon} \, 1\!\!1_h - b\left((J_x^2 - \tfrac{1}{3}J^2)\epsilon_{xx} + c.p.\right) - \frac{d}{\sqrt{3}}\left([J_x J_y]\epsilon_{xy} + c.p.\right), \qquad (4.84)$$

with deformation potentials a, b and d.

To describe the scattering of heavy holes in quantum wells, this Hamiltonian has to be projected onto the heavy hole Γ_6 subspace. This results in

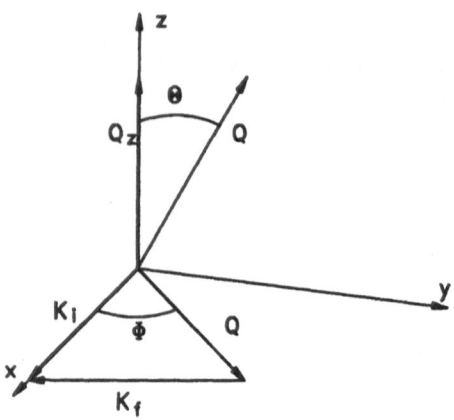

Fig. 4.6
Schematic picture of the intra–band relaxation of two–dimensional excitons by phonons. For symbols see text.

$$H_e^{hh} = \left(D_\parallel(\epsilon_{xx} + \epsilon_{yy}) + D_\perp\epsilon_{zz}\right)1\!\!1_{hh} \tag{4.85}$$

with $D_\parallel = a + b/2$ and $D_\perp = a - b$ being the effective deformation potentials for heavy hole scattering. From this interaction the deformation potential coupling constants for hole scattering can be derived as in [94] to be

$$\Xi_v^{D,LA}(Q) = \sqrt{\frac{\hbar}{2\rho N\Omega_0 u_{LA}}}|Q|^{1/2}\left(D_\parallel\cos^2(\Theta) + D_\perp\sin^2(\Theta)\right) \tag{4.86}$$

for the LA mode and

$$\Xi_v^{D,TA}(Q) = \sqrt{\frac{\hbar}{2\rho N\Omega_0 u_{TA}}}|Q|^{1/2}D_\perp\cos^2(\Theta)\sin^2(\Theta) \tag{4.87}$$

for the TA modes. Here Θ is the polar angle of the phonon wavevector Q (see Fig. 4.6)

For the spectrally discriminated scattering rate ($\sigma =$ LA,TA) the integration using the coordinate system shown in Fig. 4.6 gives the following expression:

$$\Gamma_\sigma^{S,AS}(\epsilon, \epsilon') = c_\sigma^{2d}\int_{\Theta_{min}}^{\Theta_{max}} d\cos\Theta\frac{[C^{S,AS}(Q,\Theta)]^2}{|\sin\Phi_0|}\left(n_{phon}(|\epsilon - \epsilon'|,T) + \frac{1}{2}\pm\frac{1}{2}\right) \tag{4.88}$$

whereby the phonon wavevector is given by

$$Q = \frac{|\epsilon - \epsilon'|}{\hbar u_\sigma}. \tag{4.89}$$

The quantity $C^{S,AS}$ is introduced as an abbreviation for

$$\begin{aligned}
C^{S,AS}(Q,\Theta) = \ & \left[d_c^\sigma q^{2d}(\alpha_h Q\sin\Theta, Q\cos\Theta)\right. \\
& \left. - d_v^\sigma q^{2d}(\alpha_e Q\sin\Theta, Q\cos\Theta)\right]\cdot A_\sigma(\Theta, \Phi_0)
\end{aligned} \tag{4.90}$$

with

$$d_c^{DP,LA} = \frac{D_c}{D_c - D_{\parallel}} \quad \text{and} \quad d_c^{PE,\sigma} = 1 \tag{4.91}$$

and

$$d_v^{DP,LA} = \frac{D_{\parallel} \cos^2(\Theta) + D_{\perp} \sin^2(\Theta)}{D_c - D_{\parallel}} , \quad d_v^{DP,TA} = \frac{D_{\perp} \cos^2(\Theta) \sin^2(\Theta)}{D_c - D_{\parallel}} \tag{4.92}$$

and

$$d_v^{PE,\sigma} = 1 . \tag{4.93}$$

The quantity A follows for PE–scattering from Eq. (4.66) and is given by

$$A_{LA}^2 = \frac{1}{\tilde{Q}} 9(1 - \cos^2 \Theta)^2 \cos^2 \Theta \cos^2 \Phi_0 \sin \Phi_0 \tag{4.94}$$

$$A_{TA}^2 = \frac{1}{\tilde{Q}}(1 - 3\cos^2 \Theta)^2 \sin^2 \Theta \cos^2 \Phi_0 \sin \Phi_0 \tag{4.95}$$

$$+ \sin^2 \Theta \cos^2 \Theta (2\cos^2 \Phi_0 - 1)^2 .$$

For DP–scattering we have

$$A_{LA}^2 = \tilde{Q}_{LA} \tag{4.96}$$

with $\tilde{Q}$ the dimensionless phonon wavevector

$$\tilde{Q}_{\sigma} = \frac{\hbar|\mathbf{Q}_{\parallel} + Q_z \mathbf{e}_z|}{M^* u_{\sigma}} = \frac{|\epsilon - \epsilon'|}{2\epsilon_{0,\sigma}^*} \tag{4.97}$$

with $\epsilon_{0,\sigma}^* = M^* u_{\sigma}^2/2$.

The angles $\Phi_0^{S,AS}$, $\Theta_{max}^{S,AS}$ and $\Theta_{min}^{S,AS}$ are determined by energy and momentum conservation and are given by

$$\cos^2 \Phi_0^{S,AS} = \frac{(\tilde{Q} \sin^2 \Theta \pm 2)^2}{4\tilde{K}^2 \sin^2 \Theta} \tag{4.98}$$

$$\cos^2 \Theta_{max}^{S,AS} = \max \left[0, 1 - \left(\frac{\tilde{K}}{\tilde{Q}} \left(1 + \sqrt{1 \mp 2\tilde{Q}/\tilde{K}^2} \right) \right)^2 \right] \tag{4.99}$$

$$\cos^2 \Theta_{min}^{S,AS} = 1 - \left(\frac{\tilde{K}}{\tilde{Q}} \left(1 - \sqrt{1 \mp 2\tilde{Q}/\tilde{K}^2} \right) \right)^2 \tag{4.100}$$

with

$$\tilde{K} = \sqrt{\epsilon/\epsilon_{0,\sigma}^*} \tag{4.101}$$

the dimensionless exciton wavevector. The upper sign is valid for Stokes, the lower for anti–Stokes scattering.

The factor c_σ^{2d} comprises all proportionality constants and is given for DP–scattering by ($\sigma = $ LA, TA)

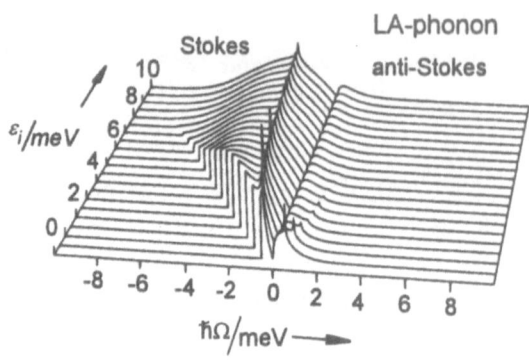

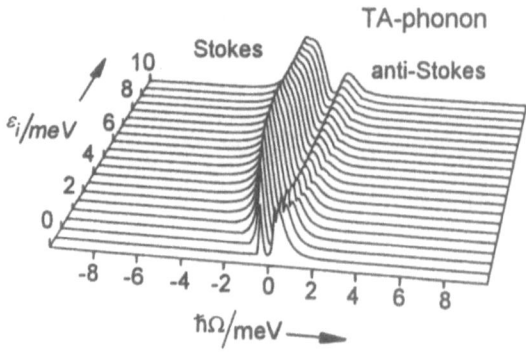

Fig. 4.7 Spectral scattering rate for exciton–phonon interaction by acoustic phonons for different exciton kinetic energies ϵ. $\hbar\Omega$: Phonon energy. $T = 7.5$ K. Upper part: DP–scattering with LA–phonons; lower part: PE–scattering with TA–phonons (vertical scale expanded by a factor 14) For other parameters see text. From [129].

$$c_{DP,\sigma}^{2d} = \frac{1}{2\pi^2} \frac{(D_c - D_{\parallel})^2 M^{*2}}{\hbar^4 \rho u_\sigma} .$$ (4.102)

For PE–scattering follows ($\sigma = $ LA, TA)

$$c_{PE,\sigma}^{2d} = \frac{2}{\pi^2} \cdot \left(\frac{ee_{14}}{\epsilon_0\varepsilon_0}\right)^2 \frac{1}{\rho\hbar^2 u_\sigma^3} .$$ (4.103)

In Fig. 4.7 the spectral scattering rates for DP–scattering by LA, and for PE–scattering by TA phonons are plotted for the following parameters: $m_{hh}^* = 0.165$, $m_e^* = 0.069$ ([121]), $L_z = 9$ nm, $a_B = 10$ nm, all other quantities have their usual value (see [127]). From these plots the essential properties of phonon scattering in two–dimensional QW can be derived. The scattering of excitons in these systems is possible down to the bottom of the band in

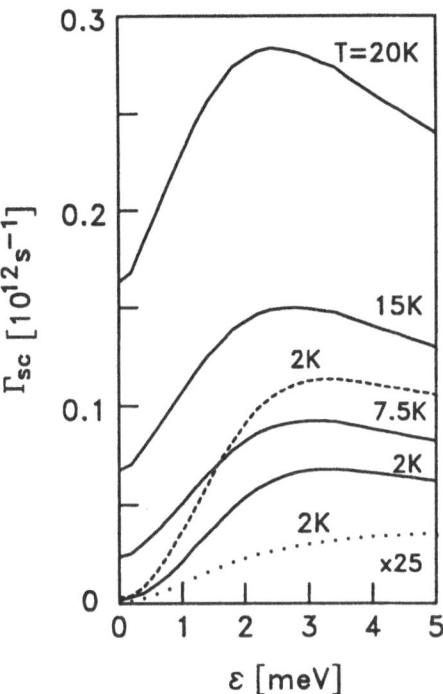

Fig. 4.8
Total scattering rate Γ_{sc} as function of exciton kinetic energy ϵ for scattering with acoustic phonons at different temperatures as indicated. DP–scattering by LA phonons: $m_{hh}^* = 0.165 m_0$ (full lines); $m_{hh}^* = 0.38 m_0$ (dashed line). PE–scattering by TA phonons: $m_{hh}^* = 0.38 m_0$ (dotted line). After [129].

contrast to the three–dimensional behaviour (compare Sect. 4.4.2.1). This is due to the non–conservation of the wavevector component perpendicular to the QW plane allowing scattering with phonons of large Q. The weak logarithmic singularities showing up as spikes in the curves can be attributed to the divergence of the one–dimensional phonon density of states perpendicular to the QW layer. These effects result in larger scattering rates of excitons in QW compared to bulk excitons with the same effective mass.

The total scattering rate following from these calculations is shown for different temperatures in Fig. 4.8. Here the mass of the heavy hole (m_{hh}) is also varied, but this has little influence on the results. In the energy range of several meV, where the calculation is quite reliable despite the assumptions made about the effective masses and dispersion relations, at low temperatures relaxation rates of the order of $5 \cdot 10^{10} \text{s}^{-1}$ corresponding to energy relaxation times of 20 ps are obtained. This corresponds to a homogeneous linewidth due to energy relaxation of about 0.03 meV, increasing linearly with temperature with a slope of $5 \mu\text{eV/K}$. These values agree quite well with those found experimentally (see Sect. 8.4).

5. Dynamics of Light Scattering from Exciton States

In order to extract the information that is contained in the time–resolved light scattering spectra an appropriate theoretical treatment is necessary. The purpose of this chapter therefore is to provide the theoretical concepts needed to describe the temporal behaviour of light scattering. We will first discuss Rayleigh scattering, which comprises all elastic light scattering processes where the system relaxes directly into the initial state after the scattering is completed. The results will then be generalized to Raman processes, which involve, in addition, the excitation of phonons in the light scattering process.

In the first step, the scattering is treated in the framework of the classical Maxwell theory for a dielectric medium. It is shown how static disorder induces the Rayleigh scattering process and which quantities determine the resonance behaviour and the temporal dependence.

In order to describe the polarization properties of the scattered light the degeneracy of the resonant intermediate states has to be included in the theory. Here we will present a simple phenomenological description of light scattering in a density matrix formalism extending the work of *Pikus* and *Ivchenko* [130] to include temporal dependence. This theory, however, does not include the requirements of the transform–limitation, i. e. both excitation and detection is treated in the white–light limit.

The excitation and detection processes themselves are included finally in the quantum mechanical theory that is developed in the last section. The treatment is based on time–dependent perturbation theory and allows one to include both phase and energy relaxation of the system with a heat bath extending previous work of *Aihara* [10]. Here analytical solutions are found, whose restriction to the white light limit reproduces the results of the density matrix treatment showing the consistency of the whole theory.

5.1 Classical Description of Resonant Rayleigh Scattering

In this section we will derive the properties of resonant Rayleigh scattering and especially its temporal dependence in the framework of classical Maxwell theory. The basic scattering experiment is sketched in Fig. 5.1. Here a plane electromagnetic wave is incident on a dielectric medium that will be characterized by a susceptibility $\chi(\mathbf{r}, t)$. The scattered wave is detected at a position $\mathbf{r}$. The incident wave creates in the medium a dielectric polarization $\mathbf{P}$. In the case of a crystal with excitonic resonances, a spatially and temporally *non–local* medium [103], the dielectric polarization is given by

$$P = 4\pi\epsilon_0 \int_V d^3 r' \int_{-\infty}^{\infty} dt' \chi(\mathbf{r}', t') E(\mathbf{r} - \mathbf{r}', t - t') , \qquad (5.1)$$

where V denotes the sample volume. For an infinite sample and stationary excitation this corresponds to a wavevector and frequency dependent susceptibility $\chi(q, \omega)$. In the following the spatially non–local properties of the medium will be neglected. This means that the theory is applicable only to localized states and all effects connected with the existence of polaritons will be disregarded. Then the frequency dependent susceptibility fulfils the equation

$$\chi_S(\mathbf{r}, t) = \int_{-\infty}^{\infty} \chi_S(\mathbf{r}, \omega) e^{-i\omega t} \, d\omega . \qquad (5.2)$$

5.1.1 The Scattered Fields

For the derivation of the scattered light the susceptibility of the medium is split into a background contribution χ_b due to electronic resonances far off the excitation frequency and a resonant part χ_S that contains the spatial fluctuations. Then the dielectric constant of the medium is given by

$$\epsilon(\mathbf{r}, t) = 1 + 4\pi(\chi_b + \chi_S) . \qquad (5.3)$$

In the following, we assume $\chi_b = 0$ which is equivalent to a background refractive index $n_b = 1$. For arbitrary n_b in all equations the velocity of light has to be replaced by c/n_b.

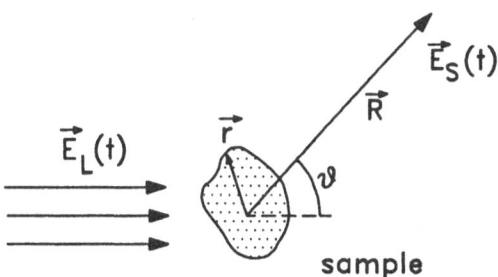

Fig. 5.1 Basic light scattering experiment to discuss the Rayleigh process. E_L: incident electric field, E_S: scattered electric field, R: detector postition, r: postition inside sample, ϑ: scattering angle. Further discussion see text.

The scattered fields are calculated by the standard procedure of scattering theory by a linearisation of Maxwell's equations treating the incident wave as the zero order solution and the scattered wave as first order solution [131]:

$$E(r,t) = E_L(r,t) + E_S(r,t) \tag{5.4}$$

with the incident field of the laser pulse

$$E_L(r,t) = A_0 \left(t - \frac{r \cdot k_L}{\omega_L} \right) e^{ik_L \cdot r - i\omega_L t} \cdot e_L . \tag{5.5}$$

The wavevector is given by $k_L = n_b \omega_L / c$ and the amplitude of the field is connected via Fresnel's equations to the wave in the exterior region of the sample. The direction of the polarization of the electromagnetic wave is denoted by e_L. The scattered field is then given by the first order Maxwell equations

$$\nabla \times E_S = -\frac{1}{c} \frac{\partial H_S}{\partial t} \tag{5.6a}$$

$$\nabla \cdot E_S = -4\pi \nabla \cdot P_S \tag{5.6b}$$

$$\nabla \times H_S = \frac{1}{c} \frac{\partial E_S}{\partial t} + \frac{4\pi}{c} \frac{\partial P_S}{\partial t} \tag{5.6c}$$

$$\nabla \cdot H_S = 0 \tag{5.6d}$$

where P_S is the polarization of the fluctuating part of the dielectric susceptibility

$$P_S(r,t) = 4\pi\epsilon_0 \int_V d^3r' \int_{-\infty}^{\infty} dt' \chi_S(r',t') E_L(r-r', t-t') . \tag{5.7}$$

These equations can be solved by introducing the Hertzian vector [131]

$$Z_S = \int_V \frac{P_S(r', t - |r - r'|/c)}{|r - r'|} d^3r' . \tag{5.8}$$

When the background dielectric constant is different from 1, the retardation has to be treated differently in the space without matter ($n_b = 1$) and those parts with matter (n_b).

The restriction to the first order solutions is equivalent to neglecting any multiple scattering processes. This means that we assume the fluctuating susceptibility χ_S to be small.

The scattered electric and magnetic fields are calculated from the Hertzian vector by

$$E = \nabla(\nabla \cdot Z) - \frac{1}{c^2} \frac{\partial^2 Z}{\partial t^2} \tag{5.9a}$$

$$H = \frac{1}{c} \nabla \times \frac{\partial Z}{\partial t} . \tag{5.9b}$$

In the following these expressions are evaluated assuming that

- the linear dimension $L = V^{1/3}$ with V the volume of the sample is small compared to the observation distance. Therefore in the spatial derivative terms proportional to $1/r^2$ and $1/r^3$ can be neglected. In the actual experiments the sample dimension is given by the diameter of the laser excitation spot;
- the time derivative is given by $\partial/\partial t \simeq -i\omega_L$, i.e. the temporal variation both of the field amplitude $A_0(t)$ and of the susceptibility χ_S is slow compared to the light frequency (SVAP approximation, see Eq. (2.2)).

Then one obtains for the scattered electric field

$$\boldsymbol{E}_S(\boldsymbol{r},t) = -k_L^2 \boldsymbol{e}_r \times [\boldsymbol{e}_r \times \boldsymbol{Z}_S(\boldsymbol{r},t)] \qquad (5.10)$$

which in the far zone is identical to the field of an electric dipole.

To calculate the Hertzian vector we restrict ourselves to materials with a spatially local response. Then it follows that

$$\boldsymbol{Z}_S = \int_V \frac{1}{|\boldsymbol{r}-\boldsymbol{r}\,'|} d^3 r\,' \int_{-\infty}^{\infty} dt' \chi_S(\boldsymbol{r}\,',t') \boldsymbol{E}_L(\boldsymbol{r}\,',t-|\boldsymbol{r}-\boldsymbol{r}\,'|/c-t'), \quad (5.11)$$

where because of causality $\chi_S(t) = 0$ for $t < 0$ has to be fulfilled.

Since $|\boldsymbol{r}| \gg |\boldsymbol{r}\,'|$ we have

$$\frac{|\boldsymbol{r}-\boldsymbol{r}\,'|}{c} \simeq \frac{r}{c} - \frac{\boldsymbol{r}\cdot\boldsymbol{r}\,'}{cr} . \qquad (5.12)$$

If further $L \ll \tau c$, where τ represents a typical time duration in which $\chi_S(t)$ or $A_0(t)$ change significantly, the last term of $|\boldsymbol{r}-\boldsymbol{r}\,'|/c$ in Eq. (5.12) can be neglected. Then we have

$$\begin{aligned} \boldsymbol{Z}_S = \; &\frac{1}{r} \int_V d^3 r\,' \int_0^{\infty} dt' \chi_s(t') A_0 \left(t - t' - \frac{r}{c} - \frac{\boldsymbol{r}\,'\cdot(\boldsymbol{k}_L - \boldsymbol{k}_S)}{\omega_L} \right) \\ &\times e^{i(\boldsymbol{k}_L - \boldsymbol{k}_S)\cdot\boldsymbol{r}\,' - i\omega_L(t - r/c) + i\omega_L t'} \cdot \boldsymbol{e}_L . \end{aligned} \qquad (5.13)$$

In this expression the scattered wavevector $\boldsymbol{k}_S$ is introduced by the definition

$$\boldsymbol{k}_S = \frac{\omega_L}{c} \boldsymbol{e}_r \qquad (5.14)$$

Exchanging the order of spatial and temporal integration leads to

$$\begin{aligned} \boldsymbol{Z}_S = \; &\frac{e^{-i\omega_L(t-r/c)}}{r} \cdot \boldsymbol{e}_L \cdot \int_{-\infty}^{\infty} e^{i\omega_L t'} \left[\int_V d^3 r\,' \chi_s(\boldsymbol{r}\,',t') \right. \\ &\left. \times A_0 \left(t - t' - \frac{r}{c} - \frac{\boldsymbol{r}\,'\cdot(\boldsymbol{k}_L - \boldsymbol{k}_S)}{\omega_L} \right) e^{i(\boldsymbol{k}_L - \boldsymbol{k}_S)\cdot\boldsymbol{r}\,'} \right] dt' . \end{aligned} \qquad (5.15)$$

Neglecting the spatial variation in $A_0(t)$ this simplifies to

$$\mathbf{Z}_S = \frac{e^{-i\omega_L(t-r/c)}}{r} \cdot \mathbf{e}_L \cdot \int_{-\infty}^{\infty} e^{i\omega_L t'} A_0(t-t'-\frac{r}{c})$$

$$\times \left[\int_V d^3\mathbf{r}' \chi_s(\mathbf{r}',t') e^{i(\mathbf{k}_L - \mathbf{k}_S)\cdot\mathbf{r}'} \right] dt' . \tag{5.16}$$

From this expression it immediately follows that the scattered field is a out-going spherical wave with mean frequency ω_L showing the elastic character of this Rayleigh scattering process. The amplitude of the scattered field is given by the *spatial Fourier transform* of the susceptibility. Rayleigh scattering therefore requires that the Fourier transform has a non–vanishing component at the scattered wavevector $\mathbf{q} = \mathbf{k}_L - \mathbf{k}_S$.

For resonant Rayleigh scattering to occur in *crystals*, the quasi–momentum selection rule, which determines the kinematics of the elastic scattering process, has to be at least partially lifted. This requires some kind of *disorder* which manifests itself as an inhomogeneous broadening of the electronic transition. Actually, resonance Rayleigh scattering in crystalline materials was first observed by *Hegarty et al.* in GaAs/AlGaAs multiple quantum–well (MQW) structures [132] where the energetically lowest ($n = 1$, e-hh) exciton exhibits appreciable inhomogeneous broadening due to compositional disorder at the interface or in the mixed crystal barrier. Using cw excitation, in their experiments these authors were able to nicely demonstrate the resonance enhancement of scattered Rayleigh intensity at the exciton transition [133, 134]. Another reason for resonant Rayleigh scattering to occur is the roughness of the surface. This effect is well known from metallic surfaces [135] and recently was also found to occur near the direct exciton resonance in bulk CdS [136, 137].

5.1.2 Monochromatic Excitation

The temporal and spectral behaviour of the scattered light is determined by the combined action of χ_S and the envelope of the exciting field A_0. The simplest case is that of a monochromatic continuous excitation given by $A_0 = const$. Introducing the frequency dependent susceptibility $\chi_S(\omega)$ in Eq. (5.16) leads to

$$\mathbf{Z}_S = \frac{e^{-i\omega_L(t-r/c)}}{r} A_0 \cdot \mathbf{e}_L \cdot \int_{-\infty}^{\infty} e^{i(\omega_L-\omega)t'}$$

$$\times \left[\int_V d^3\mathbf{r}' \chi_s(\mathbf{r}',\omega) e^{i(\mathbf{k}_L - \mathbf{k}_S)\cdot\mathbf{r}'} \right] dt' d\omega . \tag{5.17}$$

Using the identity

$$\int_{-\infty}^{\infty} e^{i(\omega-\omega')t} dt = 2\pi\delta(\omega - \omega') \tag{5.18}$$

gives for $k_S \neq k_L$ [1] as scattered wave

$$Z_S = 2\pi \frac{e^{-i\omega_L(t-r/c)}}{r} A_0 \cdot e_L \cdot \left[\int_V d^3r' \chi_s(r',\omega_L) e^{i(k_L - k_S)\cdot r'} \right] . \qquad (5.19)$$

The scattered wave has exactly the same frequency as the excitation and represents light scattered elastically from the spatially fluctuating medium. This already shows one of the main features characteristic for Rayleigh scattering, namely that the stationary scattered light has exactly the same frequency distribution as the incident light and its polarization is the same as that of the exciting laser.

For pulsed excitation the behaviour of the scattered light is more complicated. As will be shown in the next section how the susceptibility of the medium influences both the spectral and temporal properties of the scattered signal.

5.1.3 Pulsed Excitation: Spectrally Integrated Intensity

Ignoring the spectral information contained in the scattered light, the intensity at a photodetector at the position r according to Chap. 2 is given by [5]

$$\langle I(r,t) \rangle = \langle E^*(r,t) E(r,t) \rangle . \qquad (5.20)$$

Here $\langle \rangle$ denotes an ensemble average. Then the scattered intensity is

$$\begin{aligned}
I(r,t) &= \frac{k_L^4}{r^2} \cos^2 \vartheta \int_0^\infty dt' \int_0^\infty dt'' A_0^* \left(t - t' - \frac{r}{c} \right) A_0 \left(t - t'' - \frac{r}{c} \right) e^{i\omega_L(t'-t'')} \\
&\times \left[\int_V d^3r' \int_V d^3r'' \langle \chi_s^*(r',t')\chi_s(r'',t'') \rangle e^{i(k_L - k_S)\cdot(r' - r'')} \right] \quad (5.21)
\end{aligned}$$

where ϑ is the angle between r and k_L.

The correlation function of the susceptibility $\langle \chi_s^*(r',t')\chi_s(r'',t'') \rangle$ can now be divided into two parts. The first one is the average of χ_S

$$\overline{\chi_S}(t) = \frac{1}{V} \int_V d^3r' \langle \chi_S(r',t) \rangle , \qquad (5.22)$$

being a constant throughout the sample, where we have neglected any retardation effects. The second one is due to the fluctuations in χ_S and is given by

$$\langle \chi_S^*(r',t')\chi_S(r'',t'') \rangle = \langle \chi_S^*(r',t')\chi_S(r'',t'') \rangle_c + \overline{\chi_S}^*(t')\overline{\chi_S}(t'') . \qquad (5.23)$$

---

[1] In the direction $k_L = k_S$ the scattered wave interferes with the primary wave making a special treatment necessary.

To discuss the general behaviour we assume that the correlation function of the susceptibility $\langle \chi_s^* \chi_s \rangle_c$ is different from zero only within a certain correlation length ξ_c

$$\langle \chi_s^*(r',t')\chi_s(r'',t'') \rangle_c \; \simeq \; \langle \chi_s^*(r',t')\chi_s(r',t'') \rangle_c \qquad |r''-r'| < \xi_c$$

$$\text{for}$$

$$\langle \chi_s^*(r',t')\chi_s(r'',t'') \rangle_c \; \simeq \; 0 \qquad\qquad\qquad |r''-r'| > \xi_c$$

$$(5.24)$$

This correlation length is assumed to be small compared to the wavelength of light and therefore also small compared to the dimensions of the sample. Then the integral inside the brackets in Eq. (5.21) can be divided into two parts, one integrating over the mean value, the other over the correlation part.

For infinite samples the first contribution leads to a non–zero result only for $k_L = k_S$. This gives part of the transmitted signal and is not of interest here (see however the next section). The second contribution is the proper scattered light with intensity I_S. After the substitution $r'' - r' = r'''$ one spatial integration can be performed, as $\exp(i(k_L - k_S)\cdot r''') \simeq 1$. The result is

$$I_S(r,t) \;=\; \xi_c^3 \frac{k_L^4}{r^2} \cos^2 \vartheta$$

$$\times \int_0^\infty dt' \int_0^\infty dt'' \, A_0^*\left(t - t' - \frac{r}{c}\right) A_0\left(t - t'' - \frac{r}{c}\right) e^{i\omega_L(t'-t'')}$$

$$\times \int_V d^3r' \langle \chi_s^*(r',t')\chi_s(r',t'') \rangle_c . \qquad (5.25)$$

Here we see that the scattered intensity is just given by the spatial average of the susceptibility correlation function over the sample.

As a simple example we discuss the scattering in a material with susceptibility

$$\chi_S(r,\omega) = \frac{\omega_p}{\omega_0(r) - \omega - i\Gamma/2} \qquad (5.26)$$

where ω_p is a measure of the strength of the dielectric response and $\omega_0(r)$ is the fluctuating resonance frequency leading to the fluctuation in χ_S. By Fourier transforming we obtain

$$\chi_S(r,t) = -i\omega_p \Theta(t) e^{-i\omega_0 t - \Gamma t/2} . \qquad (5.27)$$

The distribution of the resonance frequency we assume to be given by a function $g_{res}(\omega_0)$. Finally Γ denotes the spatially constant homogeneous linewidth of the states. Then we have

$$\int_V d^3r' \langle \chi_s^*(r',t')\chi_s(r',t'') \rangle = \qquad (5.28)$$

$$\omega_p^2 \int_{-\infty}^\infty d\omega_0 g_{res}(\omega_0) e^{-i\omega_0(t'-t'') - \Gamma(t'+t'')/2} .$$

Approximating the real distribution function by a constant probability $1/2\Delta\omega$ around the mean value ω_0 with a width of $2\Delta\omega$ the integration of Eq. (5.28) results in

$$e^{-i\omega_0(t'-t'')-\Gamma(t'+t'')/2} \cdot \frac{\sin \Delta\omega(t'-t'')}{\Delta\omega(t'-t'')}$$

Now the two limiting situations will be discussed

1. $|\omega_L - \omega_0| \ll \Delta\omega$

   Here the first factor together with $e^{i\omega_L(t'-t'')}$ oscillates slowly compared to the second factor. If in addition $1/\Delta\omega$ is small compared to the relevant time scales of the variation of $A_0(t)$, the second factor works approximately like a $\delta(t)$–function for the time integration. Then the scattered intensity is

   $$I(\mathbf{r},t) = \xi_c^3 \omega_p^2 \frac{k_L^4}{r^2} \cos^2 \vartheta g_{\text{res}}(\omega_L) \cdot \int_0^\infty dt' |A_0(t - t' - \frac{r}{c})|^2 e^{-\Gamma t} . \quad (5.29)$$

   The temporal behaviour of the scattered intensity is determined here by a convolution of the temporal behaviour of the dielectric medium, itself determined by the homogeneous linewidth, with the temporal behaviour of the excitation intensity.

2. $|\omega_L - \omega_0| \gg \Delta\omega$

   Now $e^{i(\omega_0-\omega_L)(t'-t'')}$ oscillates fast compared to all other quantities which therefore can be replaced by their values at $t' = t'' = 0$, and the following response is obtained after performing the time integration

   $$I(\mathbf{r},t) = \xi_c^3 \frac{k_L^4}{r^2} \cos^2 \vartheta \cdot \frac{\omega_p^2}{(\omega_L - \omega_0)^2 + \Gamma^2/4} |A_0(t - \frac{r}{c})|^2 . \quad (5.30)$$

   This relation shows that far away from the resonance the scattered light instantaneously follows the excitation, while the scattered intensity shows a typical resonance profile.

With these results, a longstanding question about the time dependence of resonant Rayleigh scattering [132] is answered. In resonance the scattering process always decays with a finite time constant that has to be identified with the coherence time of the system.

To allow a more quantitative discussion of real disordered systems, the susceptibility correlation function will now be specified explicitly. The important quantities characterizing $\chi_S(\mathbf{r},\omega)$ are the distribution function of resonance frequencies denoted by $g(\omega)$ and the spatial correlation function that will be assumed to be of gaussian form

$$\langle \chi_S^*(\mathbf{r},\omega)\chi_S(\mathbf{r}',\omega') \rangle = \qquad (5.31)$$
$$\langle \chi_S^*(\mathbf{r},\omega)\chi_s(\mathbf{r},\omega') \rangle_0 \exp\left(-|\mathbf{r} - \mathbf{r}'|^2/\xi_c(\omega,\omega')^2\right) .$$

Here $\langle \chi_S^*(\boldsymbol{r}, \omega)\chi_S(\boldsymbol{r}, \omega')\rangle_0$ is the correlation function for $\boldsymbol{r} = \boldsymbol{r}'$ and $\xi_c(\omega, \omega')$ is the correlation length which is generally frequency dependent. It characterizes the spatial extent over which coherence of the electronic states exists, a finite ξ_c showing the localized nature of the states. To avoid unnecessary computational complications, in the calculations we will disregard any influence of the surface of the scattering medium and also neglect the background susceptibility.

For excitation within the inhomogeneously broadened electronic transition (distribution function $g(\omega)$ of transition frequencies) one finds

$$I(\boldsymbol{R}, t) \ \propto \ \cos^2\vartheta \cdot \xi_c^3 \cdot \exp\left(-|\boldsymbol{k}_L - \boldsymbol{k}_S|^2 \cdot \xi_c^2/4\right) g(\omega_L) \qquad (5.32)$$
$$\times \int_0^\infty dt'\,|A_0(t - \frac{R}{c} - t')|^2 e^{-(\delta E_{\text{hom}}/\hbar)\cdot t'}$$

with ϑ denoting the scattering angle between vectors $\boldsymbol{k}_L$ and $\boldsymbol{k}_S$ (Fig. 5.1). In order to derive Eq. (5.32) and to perform the necessary integration over the space coordinate, one has to presume that the correlation length ξ_c of the scattering regions is frequency independent. According to Eq. (5.32), the scattered intensity shows a pronounced dependence on the scattering angle, which sensitively reflects the spatial structure of the disorder and in principle would allow one to determine its correlation length ξ_c.

5.1.4 Two–Dimensional Systems

The relations derived in the previous section will now be specialized to two–dimensional systems. This is for two reasons.

1. In two dimensions, propagation effects can in most cases be neglected and a clear distinction between reflection and scattering is possible.
2. The calculations are directly applicable to concrete experiments in the investigations of quantum well structures (see Chap. 8).

In the following ρ denotes the two–dimensional position variable and $\boldsymbol{f}$ the vector normal to the plane. Then the wavevector of the incident and scattered light can be decomposed into a part parallel to the plane and perpendicular to it

$$\boldsymbol{k} = \boldsymbol{k}_\| + \boldsymbol{k}_\perp \quad \boldsymbol{k}_\perp = (\boldsymbol{k} \cdot \boldsymbol{f})\boldsymbol{f} \ . \qquad (5.33)$$

5.1.4.1 The Reflected Signal. The separation of the susceptibility correlation function in the two components $\overline{\chi_S^*}(t')\overline{\chi_S}(t'')$ and $\langle \chi_S^*(\boldsymbol{r}', t')\chi_S(\boldsymbol{r}'', t'')\rangle_c$ in Eq. (5.23) suggests a separate discussion. Here we consider the first contribution. Like in a homogeneous system it contains a part of the transmitted intensity ($\boldsymbol{k}_L = \boldsymbol{k}_S$) that will not be discussed further. Another non–vanishing contribution represents the reflected signal (see Fig. 5.2) that occurs for an infinitely extended system if the condition

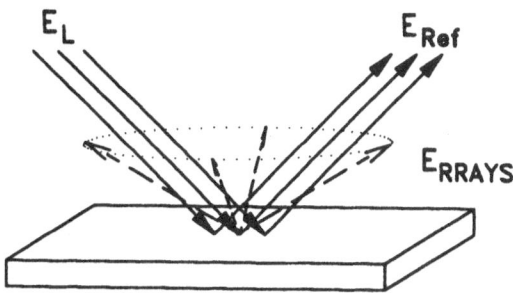

Fig. 5.2 Rayleigh scattering in two–dimensional systems. E_L denotes the exciting wave, E_{Ref} the reflected and E_{RRAYS} the scattered wave.

$$k_{\|L} - k_{\|S} = 0 \, , \tag{5.34}$$

holds, as then the argument of the exponential function vanishes. This is the usual law of reflection. For a finite system or if the excitation spot is of finite size given by a length d_0, the integration has to be performed explicitly leading to a spatial distribution of the reflected light that is the Fourier transform of the excitation distribution. This corresponds to the Fraunhofer diffraction of the excitation spot [138].

Assuming the dielectric response function again to be a Lorentz function and isotropic with respect to the polarization of the excitation[2] we obtain as temporal behaviour of the dielectric response

$$\chi_S(\rho, t) = -i\omega_p \Theta(t) \exp\left(-i\omega_0(\rho)t - \Gamma t/2\right) \, . \tag{5.35}$$

Here the spatial dependence of the resonance frequency is understood in the sense that different parts of the sample with extension smaller than the wavelength of light are characterized by a definite local transition energy. For the spectral distribution of the transition frequencies a gaussian lineshape is assumed:

$$g(\omega_0) = \frac{1}{\sqrt{\pi}\delta\omega_R} e^{-(\omega_R - \omega_0)^2/\delta\omega_R^2} \, . \tag{5.36}$$

The full spectral halfwidth, i.e. the inhomogeneous broadening, is given by $\delta E_{in} = \hbar 2\sqrt{\ln 2}\delta\omega_R$. For the sample (excitation spot) a disc–like shape with radius d_0 and area $F = \pi d_0^2$ is assumed. Under these conditions the average time dependent susceptibility can be calculated straightforwardly to be

$$\overline{\chi_S}(t) = \frac{1}{F}\int_F d^2\rho'\langle\chi_S(\rho', t)\rangle = -i\omega_p \cdot e^{-i\omega_R t - \Gamma t/2 - \frac{\delta\omega_R^2}{4}t^2} \, . \tag{5.37}$$

Insertion of this expression into Eq. (5.21) gives the temporal dependence of the reflected signal. For the calculation we assume an excitation pulse with gaussian lineshape and halfwidth $\Delta t_{1/2} = 2\sqrt{\ln 2}t_P$

---

[2]The case of degenerate states with different optical transition moment is treated in Sect. 5.3.

$$A_0(t) = A_n \frac{1}{\sqrt{\pi}t_P} e^{-(t-t_0)^2/t_P^2} . \tag{5.38}$$

The normalization factor A_n is given by the pulse energy and by the size of the excitation spot on the sample surface ($A_n \propto 1/d_0$). The time of the pulse maximum t_0 is set to zero.

As is shown by Eq. (5.23) the reflected intensity is just the square of the reflected field amplitude A_{ref} . This follows as the convolution of the susceptibility with the excitation pulse as

$$
\begin{aligned}
A_{\text{ref}}(t) &= \int_0^\infty dt' A_0(t - t') \overline{\chi_S}(t') \\
&= A_n \frac{-i\omega_p}{2\sqrt{C}} \cdot \left[1 + \text{erf}\left(\frac{t + z_{LR}t_P^2/2}{\sqrt{C}t_P} \right) \right] \\
&\quad \times \exp\left[-\frac{1}{t_p^2}\left(t^2 - \frac{(t + z_{LR}t_P^2/2)^2}{C} \right) \right] .
\end{aligned}
\tag{5.39}
$$

Here the following abbreviations were introduced

$$z_{LR} = i(\omega_L - \omega_R) - \Gamma/2 \quad C = 1 + \delta\omega_R^2 t_P^2/4$$

and $\text{erf}(z)$ is the error function [66].

This relation shows that the temporal dependence is determined both by the homogeneous and by the inhomogeneous width of the resonance. When the inhomogeneous width is small and can be neglected, the temporal decay of the reflected signal proceeds with the coherence time of the electronic state. Recently, this effect has been observed by *Aaviksoo* and *Kuhl* [139].

5.1.4.2 The Scattered Signal. From the general discussion we expect a non–vanishing intensity due to the fluctuating part of the susceptibility also in directions different from that of the reflected light. For the explicit calculation we again approximate the correlation function, the spatial variation of which is characterized by a correlation length ξ_c, by a rectangular shape that is non-zero and of constant value in a disc of radius ξ_c around an arbitrary point ρ. Then the first area integration gives a factor $\pi\xi_c^2$. The second integration can, after transformation of the area integration into an integration over the distribution of resonance frequencies, be written as

$$
\begin{aligned}
\frac{1}{F} \int_V d^2\rho' &\langle \chi_s^*(\rho', t') \chi_s(\rho', t'') \rangle_c \\
&= \frac{1}{F} \int_V d^2\rho' \langle \chi_s^*(\rho', t') \chi_s(\rho', t'') \rangle - \overline{\chi_S}^*(t') \overline{\chi_S}(t'') \tag{5.40} \\
&= \omega_p^2 \cdot \frac{\xi_c^2}{d_0^2} e^{-i\omega_R(t''-t')-\Gamma(t''+t')/2} \\
&\quad \times \left[e^{-\frac{\delta\omega_R^2}{4}(t''-t')^2} - e^{-\frac{\delta\omega_R^2}{4}(t''^2+t'^2)} \right] \tag{5.41}
\end{aligned}
$$

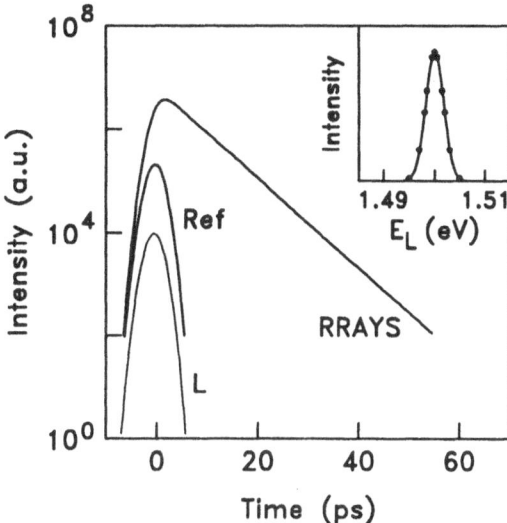

Fig. 5.3 Temporal dependence of resonant Rayleigh scattering (RRAYS) and reflection (Ref) for a inhomogeneously broadened two–dimensional medium after excitation by a short transform–limited laser pulse (L) of width $t_P = 3\,\text{ps}$ (corresponding to 5 ps FWHM). The inhomogeneous spectral width is $\delta E_{\text{in}} = 5\,\text{meV}$, the homogeneous width is $\hbar\Gamma_{\text{hom}} = 0.15\,\text{meV}$ (corresponding to $\tau_{\text{coh}} = 40\,\text{ps}$). Excitation at resonance $E_L = \hbar\omega_0$. The inset shows the dependence of the time–integrated intensity of the Rayleigh scattered light on the excitation photon energy E_L following the inhomogeneous distribution of resonance energies.

whereby the exciting laser pulse is assumed to be of gaussian shape. In the second term of the sum in Eq. (5.40) the times t' and t'' are separated allowing a straightforward integration leading to the square of the reflection term (Eq. (5.39)), but reduced by a factor $(\xi_c/d_0)^2$. In the first term of the sum the times t' and t'' arc coupled and no analytic solution of the integral is possible. Therefore we have

$$
\begin{aligned}
I_{ray}(t) \;=\; & \omega_p^2 \cdot |A_n|^2 \frac{\xi_c^2}{d_0^2}\frac{1}{\sqrt{\pi}t_P}\frac{1}{2\sqrt{C}} \\
& \times \int_0^\infty dt' \left\{ 1 + \text{erf}\left[\frac{\sqrt{C}}{t_P}\left(t' + \frac{t - t' + z_{LR}t_P^2/2}{C}\right)\right]\right\} \\
& \times \exp\left\{ -\frac{1}{t_p^2}\left[2(t - t')^2 - \frac{(t - t' + z_{LR}t_P^2/2)^2}{C} + \Gamma t' t_P^2\right]\right\} \\
& -\frac{\xi_c^2}{d_0^2}|A_{\text{ref}}(t)| \;.
\end{aligned}
\tag{5.42}
$$

The abbreviations z_{LR} and C have the same meaning as in Eq. (5.39). It is obvious from this relation that the scattered intensity is reduced by a factor

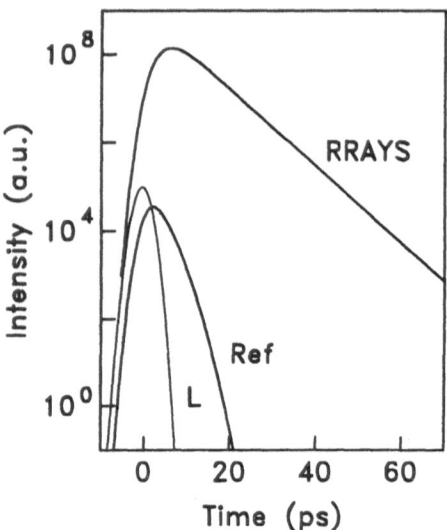

Fig. 5.4
Temporal dependence of resonant Rayleigh scattering (RRAYS) and reflection (Ref) for a inhomogeneously broadened two–dimensional medium after excitation by. a short transform–limited laser pulse (L) of width $t_P = 3\,\text{ps}$ (corresponding to 5 ps FWHM). The inhomogeneous spectral width is $\delta E_{in} = 0.25\,\text{meV}$, the homogeneous width is $\hbar\Gamma_{\text{hom}} = 0.15\,\text{meV}$ (corresponding to $\tau_{\text{coh}} = 40\,\text{ps}$). Excitation at resonance $E_L = \hbar\omega_0$.

$(\xi_c/d_0)^2$ compared to the reflected signal. This means that it is experimentally advantageous to use an excitation spot as small as possible. Furthermore the relation shows that the signal reflected from the correlation area, i.e. a disc with radius ξ_c, is missing in the scattered light.

Inspection of the integral shows that in the important case of $\delta\omega_R \gg \Gamma$ for long enough times the scattered light alwas decays with the reciprocal of the homogeneous linewidth. This explicitly verifies the discussion of the last section.

Summarizing, we see that the time dependence of the resonant Rayleigh scattering process is quite different from that of the specularly reflected intensity [139], which should decay with the reciprocal of the total energy width. Qualitatively, this is due to the difference in phase relaxation of the electromagnetic waves scattered from different points of the sample. While in specular reflection all waves add up coherently due to their identical phase, in the Rayleigh process we have a statistical distribution of phases and therefore random superposition of the scattered waves.

It is interesting to note that the ratio of the total reflected and scattered intensity is determined by the correlation length of the susceptibility fluctuations. This should allow another direct experimental determination of χ_c besides from measuring the angular distribution of the scattered light. Such experiments are at present under way.

To demonstrate these results, model calculations of the temporal dependence of the reflected and Rayleigh scattered light for different parameters are shown in Figs. 5.3 and 5.4. In both cases the correlation length was chosen as $d_0/\xi_c = 100$. In Fig. 5.3 the temporal dependence is shown for the case where the inhomogeneous broadening of the transition is much larger than the homogeneous linewidth. We see that the reflected signal follows almost instantaneously the excitation pulse since here the correlation time is extremely short (0.6 ps), being determined by the inhomogeneous width of 5 meV. In contrast, the Rayleigh scattered light exhibits a single exponential decay with a time constant of half the coherence time (20 ps) as expected. On the other hand, the time–integrated intensity for different excitation photon energies follows the distribution of transition energies (see inset in Fig. 5.3).

In the other case the homogeneous and inhomogeneous linewidth are of comparable magnitude (see Fig. 5.4). Then both the resonant reflection and Rayleigh scattering decay after short pulse excitation with a similar finite lifetime. Experimentally the expected temporal behaviour of the resonant Rayleigh scattering was demonstrated recently by our group [140] (see also Chap. 8).

The light scattering in disordered solids was recently also calculated quantum mechanically by *Aihara* [141]. His calculation also finds two components in the scattered light. One is fast and its temporal behaviour follows the excitation pulse. This component is interpreted as the disorder–induced scattering. The other part decays with the lifetime of the electronic states which in this model is identical with the coherence time and is interpreted as hot luminescence. The calculations presented here show, however, that this interpretation cannot be correct as the reflected signal also contributes to the total scattered light, a distinction from the true Rayleigh scattered signal being possible only by the different angular dependence. Since the angular dependence is not performed in the calculations the reflected signal cannot be discriminated from the Rayleigh scattering. This might be the reason for the different identifications in this work and in Ref. [141]. Furthermore the hot luminescence as a spontaneous process is not present in a classical theory that describes only the coherently driven processes.

5.2 Phenomenological Description by Density Matrix Theory

The main drawback of the classical description of resonant Rayleigh scattering (the same holds true for the classical theory of Raman scattering) turns out to be the omission of hot luminescence processes that become a strong contribution to the fluorescence under resonant excitation. In this section a systematic description of the resonance fluorescence of excitonic systems with degenerate states will be given that includes both the scattering and the hot luminescence, albeit in the white-light limit neglecting the details of the excitation and the detection process. As such it allows a qualitative understanding of light scattering from exciton states. In continuation of the treatment of the last section, we will consider only processes without any spectral shift of the scattered light which, in accordance with atomic systems [8], will be called *resonance fluorescence*.

5.2.1 The Density Matrix and Relaxation Processes

In this section, we consider an excitonic system with a single ground state $|0\rangle$, but an N–fold degenerate excited state $|i\rangle$, $i = 1, \ldots, N$, the energetic splitting of which is small compared to the distance to the ground state. The Hamiltonian of this system can be written by using excitation and annihilation operators (see e.g. [5]) $c_{0i}^{\dagger} = |i\rangle\langle 0|$ of the exciton states as

$$ H_0 = \sum_{i=0,\ldots,N} E_i c_{0i}^{\dagger} c_{0i} . \tag{5.43} $$

The state of this excitonic system is fully specified by the density matrix ρ:

$$ \rho = \sum_{i,j=0,\ldots,N} \rho_{ij} |i\rangle\langle j| \tag{5.44} $$

with components ρ_{ij}. For example, ρ is given in the frequently realized case $N = 2$ by a 3×3 matrix

$$ \rho = \begin{pmatrix} \rho_{00} & \rho_{01} & \rho_{02} \\ \rho_{10} & \rho_{11} & \rho_{12} \\ \rho_{20} & \rho_{21} & \rho_{22} \end{pmatrix} . \tag{5.45} $$

The diagonal elements of ρ describe the population of the states, while the so called *coherence* of the system is described by the non–diagonal elements. Here one can distinguish between the optical coherence ρ_{0i}, which characterize the coupling of the ground state to the excited states by interaction with the light field and the Hertzian coherence ρ_{ij} with $i, j = 1, \ldots, N$ that describe phase relations between the excited states.

The temporal evolution of the density matrix is given by a generalized master equation that is obtained from the Liouville equation by supplementing an empirical relaxation term [9] as

$$ \frac{d\rho}{dt} = -\frac{i}{\hbar} [H_0 + V, \rho] + \left(\frac{\partial \rho}{\partial t} \right)_{\text{relax}} . \tag{5.46} $$

The relaxation term describes the interaction of the system of interest with a heat bath. Assuming a fast dissipation of the interaction effects into the heat bath (Markov approximation), the relaxation term can be put in the form of a relaxation matrix $\underline{\underline{R}}$ coupling the density matrix elements:

$$\left(\frac{\partial \rho_{ij}}{\partial t}\right)_{\text{relax}} = \sum_{i',j'} R_{iji'j'} \rho_{i'j'} \ . \tag{5.47}$$

As any exact form of the relaxation matrix requires the solution of the full system and bath Hamiltonian, which of course is not possible, one often uses a simple phenomenological ansatz. This describes in the form of rate constants the energy relaxation

$$\left(\frac{\partial \rho_{ii}}{\partial t}\right)_{\text{relax}} = -\gamma_i \rho_{ii} \ , \tag{5.48}$$

which governs the population of the states, and also the phase relaxation

$$\left(\frac{\partial \rho_{ij}}{\partial t}\right)_{\text{relax}} = -\gamma_{ij} \rho_{ij} \ , \tag{5.49}$$

which destroys the coherence of the system. In most cases the phase relaxation rates are assumed in the form

$$\gamma_{ij} = \frac{1}{2}(\gamma_i + \gamma_j) + w_{ij} \ , \tag{5.50}$$

whereby w_{ij} are called the 'pure' phase relaxation rates.

This assumption can be made more physical by taking the symmetry of the states into account [130]. The states of the system are assumed to transform under the operations of the symmetry group according to the representation Γ. Then the density matrix that transforms according to the representation $\Gamma^* \otimes \Gamma$ can be decomposed into irreducible components $\Gamma^* \otimes \Gamma = \sum_\nu \Gamma^{(\nu)}$. In most cases the heat bath with which the system interacts is isotropic. Then the relaxation equation decomposes into a set of independent equations, since only density matrix components belonging to the same row i of the same representation are transformed into each other (Wigner–Eckhard theorem).

$$\left(\frac{\partial \rho_i^\nu}{\partial t}\right)_{\text{relax}} = -W^\nu \rho_i^\nu \ . \tag{5.51}$$

Because of conservation of the number of states, the sum over all relaxation rates of those components of the density matrix that transform under the identity representation $\Gamma^{(1)}$ has to vanish . Using these relations allows a great reduction of the number of relaxation rates that are necessary to describe the system and which in most cases have to be determined experimentally. Furthermore, one notices by this procedure which components of the density matrix relax in the same way.

Under certain conditions [9] the master equation is equivalent to the usual optical Bloch equations. Therefore the relaxation rates are often given in the form of the energy and phase relaxation times T_1 and T_2. For the case of $N = 2$ the relations $\gamma_\alpha = 1/T_1$, $\gamma_{0\alpha} = 1/T_2^{\text{opt}}$ and $\gamma_{12} = 2/\tau_{\text{coh}}$ hold, whereby

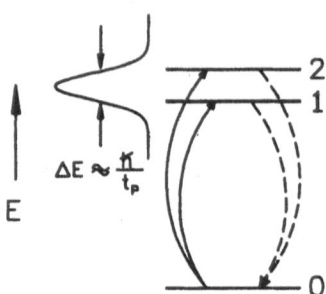

Fig. 5.5
Schematic three level system with ground state $|0\rangle$ and nearly degenerate excited states $|1\rangle$ and $|2\rangle$. The full arrows denote absorption and spontaneous emission processes. The drawing to the left shows the spectral intensity distribution of the coherent laser pulse with duration t_p and spectral width $\Delta E \simeq \hbar/t_p$.

the factor of two in the last relation allows a direct comparison of the optical dephasing time (T_2) with the quantum dephasing or coherence time (τ_{coh}).

When only energy relaxation occurs, from Eq. (5.50) one notes that the simple relation $T_2^{\mathrm{opt}} = \tau_{\mathrm{coh}} = 2T_1$ is valid [7]. In the general case no information exists about the different relaxation rates which have to be determined experimentally.

5.2.2 Interaction with the Light Field

We now allow our system to interact with the light field. We thereby confine ourselves to the dipole interaction and consider an electromagnetic wave with a well defined polarization vector e_L. The temporal variation of the field we write down in the SVAP approximation as $E^{(+)}(t) = F_L(t)\exp(-i\omega_L t)$, corresponding to the positive frequency part of the wave. Then we have for $N = 2$

$$V = \begin{pmatrix} 0 & M_1^* & M_2^* \\ M_1 & 0 & 0 \\ M_2 & 0 & 0 \end{pmatrix} \cdot (E^{(+)}(t)e_L + c.c.) . \tag{5.52}$$

Here M_i is the vector transition moment of the state $|i\rangle$.

$$M_i = \langle i|\, D\, |0\rangle \quad i = 1,\ldots,N , \tag{5.53}$$

and D is the dipole operator.

The exciting laser pulse has a certain duration t_P and because of the time–frequency uncertainty also a corresponding spectral width $\Delta E \simeq \hbar/t_P$, as shown schematically on the left of Fig. 5.5. Therefore the laser pulse can excite energetically split states if the splitting is small compared to the laser spectral width.

The scattered light that is emitted from the excitonic system can be described by an expression similar to that derived in the classical theory of the foregoing section ('source field representation') as was first discussed by *Lemberg* [110] (see also [5]). Here the electric field operator is given in the Heisenberg picture of quantum mechanics under the same assumptions that were introduced in the classical treatment, in detail

- the spatial extent of the radiating system is small compared to the distance to the detector,
- the temporal variation occurs on a time scale long compared to the reciprocal transition frequency,
- a direct overlap of the scattered light and the excitation is excluded.

This means in particular that all propagation effects (polariton effects) are not included and the light intensity is small.

Then the scattered light with polarization vector e_S is given by

$$E_S^{(+)}(t) \propto \sum_{i=1,\dots,N} (M_i \cdot e_S)^* c_{0i}(t - R/c) , \qquad (5.54)$$

where R denotes the distance from the sample to the detector. In most cases, however, the retardation, i.e. $R \neq 0$, can be neglected. The intensity of the scattered light is given according to Chap. 2 by the correlation function of the electric field strength

$$\langle\, E_S^{(-)}(t)\, E_S^{(+)}(t') \,\rangle \propto \sum_{i,j=1,\dots,N} (M_i \cdot e_S)^* (M_j \cdot e_S)^* \langle\, c_{0i}^\dagger(t)\, c_{0j}(t') \,\rangle . \qquad (5.55)$$

Here the average is defined in the usual sense as the trace with the exciton density matrix. The calculation therefore requires the evaluation of the two–time correlation function. This will be postponed until Sect. 5.3.

5.2.3 Broad–Band Excitation and Detection

In this section we calculate the temporal dependence of the scattered light in the limit of broad–band detection, the 'white light' limit of atomic fluorescence theory. The spectral filter then has a simple $\delta(t)$–function like response. According to Eq. (2.14) the intensity of the scattered light is given by the correlation function at equal times $t = t'$. The equal–time correlation function $\langle\, c_{0i}^\dagger(t)\, c_{0j}(t) \,\rangle$ is given by the corresponding density matrix elements in the Schrödinger picture

$$\langle\, c_{0i}^\dagger(t)\, c_{0j}(t) \,\rangle = \rho_{ij}(t) . \qquad (5.56)$$

The scattered intensity for the simple three–level system is obtained as

$$I(t) \propto |M_1 \cdot e_S|^2 \rho_{11}(t) + |M_2 \cdot e_S|^2 \rho_{22}(t) + 2\Re e\,[(M_1 \cdot e_S)^*(M_2 \cdot e_S)\rho_{12}(t)] . \qquad (5.57)$$

In the case of N–fold degenerate excited states this can be written

$$I(t) \propto \sum_{i,j=1}^{N} (M_i \cdot e_S)^*(M_j \cdot e_S)\rho_{ij}(t) \qquad (5.58)$$

where the ground state is excluded in the summation. This shows that the scattered light is fully determined by the populations and the Hertzian coherence and not influenced by the optical coherence.

The next step is the determination of the temporal development of the density matrix. Since in most cases the relvant time scale is large compared to the reciprocal optical transition frequency, to further simplify the treatment the 'rotating wave approximation' can be made in the interaction picture. Here all terms that oscillate with optical frequencies are negelcted in the master equation. Then the following equations for the components of the density matrix are obtained

$$\frac{d\rho_{0i}}{dt} = -\frac{i}{\hbar}\left[M_i^{L*}F_L(t)(\rho_{ii}-\rho_{00}) + \sum_{j=1}^{N}M_j^{L*}F_L(t)\rho_{ji}\right] \qquad (5.59a)$$
$$-[i(\omega_i-\omega_L)+\gamma_{0i}]\rho_{0i}$$

$$\frac{d\rho_{ii}}{dt} = -\frac{i}{\hbar}\left[M_i^{L}F_L(t)^*\rho_{0i} - M_i^{L*}F_L(t)\rho_{i0}\right] - \gamma_i\rho_{ii} \qquad (5.59b)$$

$$\frac{d\rho_{ij}}{dt} = -\frac{i}{\hbar}\left[M_i^{L}F_L(t)^*\rho_{0j} - M_j^{L*}F_L(t)\rho_{i0}\right] \qquad (5.59c)$$
$$-[i(\omega_i-\omega_j)+\gamma_{ij}]\rho_{ij} .$$

The last equation describes the temporal dependence of the Hertzian coherence, i.e. for $i,j = 1,\ldots,N$ and $i \neq j$. As abbreviations $\omega_i = E_i/\hbar$ and $M_i^L = M_i \cdot e_L$ have been introduced.

This coupled inhomogeneous system of differential equations can be solved in most cases only numerically. For weak excitation an iterative procedure is possible [7]. To this end the density matrix is expanded in a series

$$\rho = \sum_{n=0}^{\infty}\rho^{(n)} \qquad (5.60)$$

where the n-th term $\rho^{(n)}$ is proportional to the perturbation $V(t)$. By the iteration relation

$$\frac{d\rho^{(n+1)}}{dt} = -\frac{i}{\hbar}\left\{\left[H_0,\rho^{(n+1)}\right] + \left[V,\rho^{(n)}\right]\right\} + \left(\frac{\partial\rho^{(n+1)}}{\partial t}\right)_{relax} \qquad (5.61)$$

it is possible to obtain systematically approximate solutions of the master equation.

As an example we consider the case of a weak instantaneous excitation with a $\delta(t)$–function like light pulse. Such a pulse is obtained for example as the limit of a series of gaussian pulses, the temporal duration of which approaches zero while their energy content, i.e. the time–integrated intensity remains finite. Then the zeroth approximation is given by

$$\rho_{00}^{(0)} = 1 \qquad (5.62)$$
$$\rho_{ij}^{(0)} = 0 .$$

In the first approximation only the optical coherences are excited

$$\rho_{0i}^{(1)}(t) = \Theta(t)\frac{i}{\hbar}M_i^L e^{-[i(\omega_i - \omega_L) + \gamma_{0i}]t} \tag{5.63}$$

with the Heavyside step function

$$\Theta(t) = \begin{cases} 0 & t < 0 \\ 1/2 & t = 0 \\ 1 & t > 0 \end{cases}. \tag{5.64}$$

Finite populations and Hertzian coherence are obtained in the second approximation, where for $i, j = 1, \ldots, N$

$$\frac{d\rho_{ij}^{(2)}}{dt} = \frac{1}{\hbar^2}M_i^L M_j^{L*}\delta(t) - [i(\omega_i - \omega_j) + \gamma_{ij}]\rho_{ij}. \tag{5.65}$$

For a three–level system solving this system of differential equations results in the following time dependence of the scattered light intensity

$$\begin{aligned} I(t) \propto\ & |M_1{\cdot}e_S|^2|M_1{\cdot}e_L|^2 \exp(-\gamma_1 t) + |M_2{\cdot}e_S|^2|M_2{\cdot}e_L|^2 \exp(-\gamma_2 t) \\ & + \Re e\left[(M_1{\cdot}e_S)^*(M_1{\cdot}e_L)(M_2{\cdot}e_S)(M_2{\cdot}e_L)^* \exp\left(-i(E_1 - E_2)/\hbar - i\gamma_{12})t\right)\right]. \end{aligned} \tag{5.66}$$

Inspecting this result we see that the emitted light is composed of two parts, a background signal decaying with the energy relaxation time and an oscillating component, the *quantum beat* [142]. This part decays with the excited state coherence time. This result is identical to the usual treatment of quantum beats in the resonance fluorescence of an atomic system [142]. Here the fluorescence of an ensemble of atoms in a coherent superposition of degenerate states, excited by a short laser pulse, is observed, thereby assuming independent absorption and emission processes. In the white–light limit the density matrix treatment therefore is equivalent to the simple picture of *absorption followed by emission* of resonant light scattering (see e.g. [51]). This relation, however, does not hold for the transform–limited theory presented in the next section.

To discuss the results we will consider a simple model system with two degenerate excited states ($N = 2$) and with orthogonal transition moments. The polarization of the excitation e_L is chosen in such a way that the relation

$$e_L{\cdot}\hat{M}_1 = 1 \quad e_L{\cdot}\hat{M}_2 = 0 \tag{5.67}$$

holds. For the scattered light the parallel polarization direction $e_S^{\parallel}$ is defined as

$$e_S^{\parallel} = e_L \tag{5.68}$$

while the orthogonal polarization $e_S^{\perp}$ is fixed by the condition

$$e_S^{\perp *} \cdot e_L = 0 \ . \tag{5.69}$$

From this it follows that

$$e_S^{\perp} \cdot \hat{M}_1 = 0 \qquad e_S^{\perp} \cdot \hat{M}_2 = 1 \ . \tag{5.70}$$

The intensity of the scattered light for the different polarization direction follows as

$$\begin{aligned}
I(e_S^{\|}) &= e^{-\gamma_1 t} & (5.71\text{a}) \\
I(e_S^{\perp}) &= 0 \ . & (5.71\text{b})
\end{aligned}$$

From this relation one would only expect non–zero intensity for emitted light parallel to the excitation polarization. This differs from the result obtained below from the generalized density matrix treatment. The reason is that the relaxation matrix γ_i, γ_{ij}, on which Eq. (5.71a) is based, does not contain a cross–relaxation term between the degenerate states. Therefore a method to obtain the most general form of the relaxation matrix is necessary and is developed in the next section.

5.2.4 Generalization for Arbitrary Light Polarization

The relation (5.65) together with Eq. (5.58) shows that for instantaneous excitation the scattered intensity is completely determined by the populations and the Hertzian coherence and that the optical coherence need not to be considered. Therefore one can restrict the treatment to the dynamics of the excited states, i.e. to the excited state density matrix

$$\rho^e = \sum_{i,j=1,\ldots,N} \rho_{ij}^e \, |i\rangle \, \langle j| \ . \tag{5.72}$$

In the master equation of this excited state density matrix ρ^e the optical excitation from the ground state can be taken into account in the form of a generation matrix

$$G_{ij}(t) = \frac{1}{\hbar^2} M_i^L M_j^{L*} \delta(t) \ . \tag{5.73}$$

With this definition it is shown that the treatment of optical orientation of an exciton system under stationary conditions developed by *Ivchenko* [130] can be extended to time–dependent problems under the condition of instantaneous excitation and broad–band detection.

It is possible to generalize the relations obtained to the case of arbitrary light polarization using the concept of the light coherence matrix. In this way a useful procedure is obtained by which, even for exciton systems showing a complicated splitting pattern and optical transition moments, the temporal and polarization dependence of the resonance fluorescence can be derived.

An arbitrarily polarized exciting laser pulse can be specified by its coherence matrix [138]. To introduce it, one defines two orthogonal polarization vectors that are described in Dirac's notation as polarization basis states

$$|1\rangle \quad |2\rangle \tag{5.74}$$

and may be chosen e.g. as unit vectors parallel to the x and y direction of a cartesian coordinate system. Except in the case of the exact $0°$ scattering geometry the basis states have to be chosen differently for excitation and detection.

The the polarization state can be specified by the polarization density matrix or the coherence matrix $\underline{\underline{I}}$

$$\underline{\underline{I}} = \sum_{\alpha,\beta} I_{\alpha\beta} |\alpha\rangle \langle\beta| \quad \alpha, \beta = 1, 2 \ . \tag{5.75}$$

The intensity of the light field $I(e)$ with polarization direction e is given by

$$I(e) = \mathrm{Tr}[\underline{\underline{I}} |e\rangle \langle e|] \ , \tag{5.76}$$

where $|e\rangle \langle e|$ is the polarization state corresponding to the unit vector e.

The polarization density matrix is closely related to the Stokes vector $S = (I_0, I_1, I_2, I_3)$ [138], through the relations

$$\begin{align}
I_0 &= I(x) + I(y) \tag{5.77a}\\
I_1 &= I(x) - I(y) \tag{5.77b}\\
I_2 &= I(x') - I(y') \tag{5.77c}\\
I_3 &= I(\sigma_+) - I(\sigma_-) \ . \tag{5.77d}
\end{align}$$

Here the states

$$|x'\rangle = \frac{1}{\sqrt{2}} (|x\rangle + |y\rangle) \tag{5.78a}$$

$$|y'\rangle = \frac{1}{\sqrt{2}} (|x\rangle - |y\rangle) \tag{5.78b}$$

$$|\sigma_\pm\rangle = \frac{1}{\sqrt{2}} (|x\rangle \pm i |y\rangle) \tag{5.78c}$$

define the polarization states for light linearly polarized with an angle of $45°$ and $135°$ referred to the x axis and for left and right circularly polarized light. Then the relation between the components of the Stokes vector and the coherence matrix is

$$\underline{\underline{I}} = \frac{1}{2} \begin{pmatrix} I_0 + I_1 & I_2 + iI_3 \\ I_2 - iI_3 & I_0 - I_1 \end{pmatrix} \ . \tag{5.79}$$

Table 5.1 Overview showing possible two–dimensional representations of various symmetry point groups of solids and the decomposition of their density matrix into irreducible components.

group	representation	density matrix
D_{2d}	Γ_5	$\Gamma_1 + \Gamma_2 + \Gamma_3 + \Gamma_4$
D_{4h}	Γ_5^-	$\Gamma_1^+ + \Gamma_2^+ + \Gamma_3^+ + \Gamma_4^+$
C_{3v}	Γ_3	$\Gamma_1 + \Gamma_2 + \Gamma_3$
D_{3d}	Γ_3^-	$\Gamma_1^+ + \Gamma_2^+ + \Gamma_3^+$
C_{6v}	Γ_5	$\Gamma_1 + \Gamma_2 + \Gamma_6$

The coupling of the exciton states with the light field in the dipole approximation can be described by two $2 \times N$ matrices $\underline{\underline{M}}^{L,S}$ for excitation and detection

$$M_{\alpha i}^L = \boldsymbol{M}_i \cdot \boldsymbol{e}_{\alpha,L} \tag{5.80}$$

with the polarization basis states for excitation $\boldsymbol{e}_{\alpha,L}$ and

$$M_{\alpha i}^S = \boldsymbol{M}_i \cdot \boldsymbol{e}_{\alpha,S} \tag{5.81}$$

with the corresponding polarization states for detection $\boldsymbol{e}_{\alpha,S}$.

Then the generation matrix is

$$\underline{\underline{G}} = (\underline{\underline{M}}^L)^\dagger \cdot \underline{\underline{I}}^L \cdot \underline{\underline{M}}^L \tag{5.82}$$

and the scattered intensity is connected to the exciton excited state density matrix by

$$\underline{\underline{I}}^S = \underline{\underline{M}}^S \cdot \rho^e \cdot (\underline{\underline{M}}^S)^\dagger . \tag{5.83}$$

Together with the equation of motion of ρ^e (the excited state *Master equation*)

$$\frac{d\rho^e}{dt} = -\frac{i}{\hbar}[H_{ex}, \rho^e] + \underline{\underline{G}} + \left(\frac{\partial \rho^e}{\partial t}\right)_{\text{spin–relax}} \tag{5.84}$$

from these relations the temporal behaviour of the scattered light can be calculated for arbitrary polarization of excitation and detection. In Eq. (5.84) H_{ex} denotes the Hamiltonian of the exciton states describing the excited state energy splittings. The relaxation matrix contains only scattering within the excited state manifold, the so–called spin relaxation [130].

Considering again the model with two degenerate excited states we now have to take the symmetry into account. Possible symmetries of two–dimensional representations are given in Table 5.1. One important choice (see Chap. 8) in the symmetry group D_{2d} is the representation Γ_5.

From the symmetry of the states the irreducible components of the density matrix and the possible relaxation processes can be found. For groups containing a three–fold rotation axis three relaxation rates are needed to fully describe the relaxation channels while for the groups D_{2d} and D_{4h} four rates are necessary.

As an example we consider the case of the Γ_5 representation in D_{2d}. Here the irreducible density matrix components are given by

$$
\begin{aligned}
\rho^{(1)} &= (\rho_{11} + \rho_{22})/2 \\
\rho^{(2)} &= (\rho_{12} - \rho_{21})/2 \\
\rho^{(3)} &= (\rho_{11} - \rho_{22})/2 \\
\rho^{(4)} &= (\rho_{12} + \rho_{21})/2
\end{aligned}
\tag{5.85}
$$

with corresponding relaxation rates γ_i. Now the density matrix can be written as

$$
\rho^e = \begin{pmatrix} \rho^{(1)} + \rho^{(3)} & \rho^{(4)} + \rho^{(2)} \\ \rho^{(4)} - \rho^{(2)} & \rho^{(1)} - \rho^{(3)} \end{pmatrix}.
\tag{5.86}
$$

The polarization basis now is chosen in such a way as to diagonalize both the excitation and the detection coupling matrix. In all cases shown in Table 5.1 these are the x and y polarized states. This means that

$$
\underline{\underline{M}}^{L,S} = \begin{pmatrix} 1 & 0 \\ 0 & 1 \end{pmatrix}.
\tag{5.87}
$$

Then the following results for the coherence matrix of the scattered light are obtained for different excitation polarizations (given by the corresponding coherence matrix)

$$
\underline{\underline{I}}^L = \begin{pmatrix} 1 & 0 \\ 0 & 0 \end{pmatrix} \qquad \underline{\underline{I}}^S = \begin{pmatrix} e^{-\gamma_1 t} + e^{-\gamma_3 t} & 0 \\ 0 & e^{-\gamma_1 t} - e^{-\gamma_3 t} \end{pmatrix}
\tag{5.88}
$$

$$
\underline{\underline{I}}^L = \frac{1}{2}\begin{pmatrix} 1 & 1 \\ 1 & 1 \end{pmatrix} \qquad \underline{\underline{I}}^S = \begin{pmatrix} e^{-\gamma_1 t} & e^{-\gamma_4 t} \\ e^{-\gamma_4 t} & e^{-\gamma_1 t} \end{pmatrix}
\tag{5.89}
$$

$$
\underline{\underline{I}}^L = \frac{1}{2}\begin{pmatrix} 1 & i \\ -i & 1 \end{pmatrix} \qquad \underline{\underline{I}}^S = \begin{pmatrix} e^{-\gamma_1 t} & ie^{-\gamma_2 t} \\ -ie^{-\gamma_2 t} & e^{-\gamma_1 t} \end{pmatrix}
\tag{5.90}
$$

These relations show that by choosing the appropriate excitation and detection polarizations all relaxation rates can be measured.

The first row leads e.g. to the following polarized intensities

$$
I(e_S^{\parallel}) = \frac{1}{2}\left(e^{-\gamma_1 t} + e^{-\gamma_3 t}\right)
\tag{5.91a}
$$

$$
I(e_S^{\perp}) = \frac{1}{2}\left(e^{-\gamma_1 t} - e^{-\gamma_3 t}\right).
\tag{5.91b}
$$

These relations are obviously different from that of the last section. This shows the importance of the choice of the relaxation matrix as here we have included all possible relaxation channels. Up to now, the theory allows one to describe the case of a multi–level system with only one ground state. In this form it is applicable to localized exciton states in semiconductors that are bound by defects or in disorder–induced fluctuating potentials. These excitons closely resemble atomic systems whereby the degeneracy and energy splittings are determined by the symmetry of the valence and conduction bands from which the excitons are formed.

5.2.5 The Case of a Degenerate Ground State

The extension of the foregoing description to systems with a degenerate ground state is straightforward and completely analogous to the treatment in the case of atomic systems (see e.g. [142]). Denoting the various ground state levels by $|l\rangle$ with $l = 1, \ldots, M$ the excitation operators are now given by the relation $c_{l,i}^{\dagger} = |i\rangle \langle l|$. In the calculation of the scattered field in Eq. (5.55) the summation has to be performed in addition over the ground state levels. In the field correlation function for the scattered intensity, however, all contributions of the form $\langle\, c_{li}^{\dagger}(t)\, c_{mj}(t')\, \rangle$ with $l \neq m$ give zero contribution because of the orthogonality of the ground state levels. In the scattered intensity (Eq. (5.56)) this leads to an additional summation of the intensities of the transitions from each ground state into all excited states according to

$$I(t) \propto \sum_{l=1}^{M} \sum_{i,j=1}^{N} (M_{il}\cdot e_S)^{*}(M_{jl}\cdot e_S)\rho_{ij}(t) \ . \tag{5.92}$$

Here the matrix element of the dipole operator is given by

$$M_{il} = <\, i|D|l\, > \quad i = 1, \ldots, N, \quad l = 1, \ldots, M \ . \tag{5.93}$$

Accordingly, the generation matrix can be written

$$\underline{\underline{G}} = \sum_{l,m=1}^{M} (\underline{\underline{M}}_{l}^{L})^{\dagger} \cdot \underline{\underline{I}}^{L} \cdot \underline{\underline{M}}_{m}^{L} \rho_{lm}^{g} \ . \tag{5.94}$$

Here ρ^{g} denotes the density matrix of the ground state before the optical excitation. The polarization tensor of the scattered light is given by

$$\underline{\underline{I}}^{S} = \sum_{l=1}^{M} \underline{\underline{M}}_{l}^{S} \cdot \rho^{e} \cdot (\underline{\underline{M}}_{l}^{S})^{\dagger} \ . \tag{5.95}$$

The coupling matrices are

$$M_{\alpha i,l}^{L} = M_{il} \cdot e_{\alpha,L} \tag{5.96}$$

with the excitation polarization basis $e_{\alpha,L}$ and

$$M^{S}_{\alpha i,l} = M_{il} \cdot e_{\alpha,s} \qquad (5.97)$$

with the detection polarization basis $e_{\alpha,s}$.

With these relations the light scattering for all possible cases of localized exciton states can be described.

The ground state density matrix ρ^g is given in most cases by a thermal distribution and the generation matrix resembles the different occupation of the states according to the Boltzmann distribution. In principle, however, there is the possibility to change the ground state density matrix by irradiation of coherent microwave radiation in analogy to optical pumping (see e.g. [143]) and to induce new coherent effects in the light scattering.

The incoherent summation over the different ground state levels expresses explicitly a fundamental principle of quantum theory. The occurrence of Hertzian coherence in the scattered light is finally due to the fact that for spectrally broad–band detection no information is available about the detailed scattering pathway. Therefore the probability amplitudes for the different scattering channels have to be summed and interference terms occur in the probability. In this respect the Hertzian coherence is analogous to a multiple slit interference experiment. In contrast, for the case of scattering into different final states, it is possible to decide after the scattering event the final state of the scattering process and all interference effects have to vanish. The observation of quantum beats for excitons is therefore also of fundamental interest in quantum mechanics.

5.2.6 Rayleigh and Raman Scattering

These considerations can also be applied to light scattering processes with the participation of phonons; these are most important for the case of light scattering from free excitons as observed in this work. In contrast to localized excitons, the free states have translational degrees of freedom based on their character of Bloch waves extending throughout the whole crystal. As we have seen in Chap. 4, the translation can be characterized by a wavevector K. In Fig. 5.6 two such states with wavevectors K_1 and K_2 are shown schematically. Therefore for elastic Rayleigh scattering with the photon energy of the scattered light all contributions are to be added coherently. As was shown in the last section this leads, in ideal two–dimensional systems, to the resonant reflection of the exciting light, while in systems with spatial disorder *transient resonant Rayleigh scattering* occurs.

In the case of *resonant Raman scattering* by phonons, however, the final state of the scattering process involves excitation of phonons in the crystal lattice, as shown schematically in Fig. 5.6 by the dashed lines. In most cases, in particular those of interest in this work, the quasi–momentum conservation in the scattering process leads to a coupling of exciton states with different K to a different phonon state. This situation corresponds exactly to the case

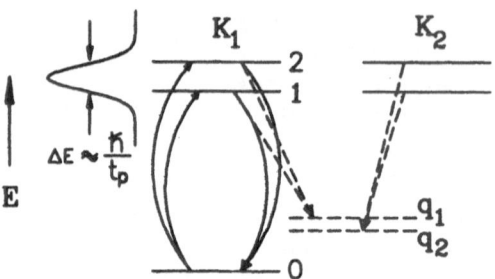

Fig. 5.6 Schematic picture of an excitonic three–level system with a ground state $|0\rangle$ and nearly degenerate states $|1\rangle$ and $|2\rangle$ representing different Bloch waves. The full arrows denote the absorption and the sponantaneous emission processes, dashed arrows Raman transitions with different phonons (the phonon states are marked by different dashed lines). For the left–hand side see Fig. 5.5.

of a degenerate ground state discussed above. Therefore we can draw the important conclusion that Raman processes involving excitons with different wavevectors do not interfere in the spontaneous scattering process, where the phonons are in thermal equilibrium[3]. If we observe quantum beats in resonant Raman processes of excitons, we can conclude that they are always due to nearly degenerate states at the same wavevector K.

5.3 Quantum Theory of Transform–Limited Light Scattering

The phenomenological description of light scattering outlined in the last section has two drawbacks rendering the quantitative analysis of experimental data difficult. One is that the relaxation processes are introduced in an 'ad-hoc' way, so that their physical interpretation is not clear. The other and even more serious problem is the assumption of broad–band excitation and detection, as experimentally the case of transform–limited excitation and registration of the scattered light is the central topic of this work.

The results of the last section nevertheless allow a qualitative understanding that is useful for a first look at the principal phenomena. For a quantitative interpretation, however, the transform–limited excitation and detection have to be incorporated into the theory from the start. Furthermore the relaxation processes should be taken into account in a way as simple as possible in order to gain an understanding of how energy and phase relaxation processes can be observed in light scattering.

---

[3]For scattering processes where the phonons are coherently excited, this might be different and one can observe here quantum beats between the phonon states (see e.g. [35]).

While the latter topic is discussed extensively in the literature (see e.g. [10, 144]), the incorporation of the excitation and detection is only at the beginning and only a few concepts exist [17–19]. The reason for this is of course that the experimental advances in time–resolved spectroscopy have only recently allowed one to reach the limit posed by the time–frequency uncertainty relation in the picosecond time range.

From the discussion in Chap. 2 it follows that the central quantity that determines the time–resolved spectrum is the field correlation function, which can be calculated by a standard perturbation procedure using Feynman diagrams to systematically take all contributions to the light scattering into account. This method was developed for stationary scattering by the group of *Toyozawa* in the seventies [145, 146] and later also applied to time–resolved situations which, however, were treated in the white–light limit [10]. These calculations are extended in this section to a transform–limited description of light scattering and also allow one to include Raman processes involving indirect excitons.

5.3.1 Outline of the Theory

In close analogy to *Aihara* [10] we consider a system that is composed of excitons, photons and a heat bath providing the relaxation processes of the exciton states. The exciton states are characterized according to Chap. 4 by their wavevector K, describing the centre–of–mass motion, and by an index $\alpha = 1, \ldots, N$, that counts the different internal exciton states, as schematically depicted in Fig. 5.5. In the limit of low exciton density, the excitons with different K and α form a system of independent oscillators and fulfil Bose commutation relations.

The photonic system comprises both the modes of the exciting laser pulse and the modes of the whole detection setup which in the usual experimental configuration are quite different (laser: gaussian beam, detector: radial waves) but here will be assumed to be plane waves characterized by wavevectors q and polarization directions e_λ.

The bath is modeled by a system of harmonic oscillators in thermal equilibrium. One example that is assumed as a model in the following would consist of the system of lattice vibrations (phonons), but at low excitation density excitonic defect states can also be described in that way. One of the aims of this section is to clarify the influence of elastic interaction with the bath states on the light scattering. Therefore the exciton–bath interaction is assumed to be of bilinear form in both the exciton and phonon states allowing the simplest description of such elastic relaxation processes [10, 146]. This bilinear form is also plausible for scattering of excitons by defects, which proceeds via free exciton annihilation at the defect side under creation of a bound state followed by the reversed process.

The Hamiltonian of the total system is then given by (in this section we set $\hbar = 1$):

$$H = H_0 + V \tag{5.98}$$

with

$$H_0 = \sum_{K,\alpha} E_{eK,\alpha} B_{K,\alpha}^\dagger B_{K,\alpha} + \sum_{q,\lambda} \omega_q a_{q,\lambda}^\dagger a_{q,\lambda} + \sum_l \epsilon_l b_l^\dagger b_l \tag{5.99a}$$

$$V = V_{AR} + V_{AB} \tag{5.99b}$$

$$= \sum_{q,\lambda} \sum_{K,\alpha} \left[g_{K,\alpha}(q,\lambda) B_{K,\alpha}^\dagger a_{q,\lambda} + g_{K,\alpha}^*(q,\lambda) a_{q,\lambda}^\dagger b_{K,\alpha} \right] \tag{5.99c}$$

$$+ \sum_{l,l'} \sum_{K,\alpha,K',\alpha'} h_{K,\alpha,K',\alpha'}(l,l') B_{K,\alpha}^\dagger b_l^\dagger b_{l'} B_{K',\alpha'} . \tag{5.99d}$$

H_0 comprises the Hamiltonians of the excitonic system, of the photons and of the bath states. The part V contains the interactions of the excitons with the light field, whereby we have included from Eq. (4.23) only the terms of the 'rotating wave' approximation and the interaction with the bath states in a bilinear form. In principle, it is possible to include in the Hamiltonian also other exciton–phonon scattering processes that allow for energy relaxation (compare Sect. 4.4). As this, however, would complicate the following calculations unnecessarily, it will not be undertaken here.

The quantities in H have the following meaning: $B_{K,\alpha}^\dagger$ denote the creation operators of exciton states with energy $E_{K,\alpha}$; $a_{q,\lambda}^\dagger$ are the creation operators of photons with wavevector q, frequency ω_q and polarization state λ. Here one has to distinguish the photons of the exciting laser pulse (wavevector q_L, frequency ω_L and polarization vector e_L) from those of the scattered light (wavevector q_S, frequency ω_S and polarization vector e_S) which are assumed as plane waves existing in different regions of space.

$b_l^\dagger$ is the creation operator of a bath state l with energy ϵ_l. The matrix elements $h_{K,\alpha,K',\alpha'}(l,l')$ describe the exciton bath scattering and are assumed to be real quantities.

Finally, the interaction of the excitons with the light field is parameterized by the coupling coefficients $g_{K,\alpha}(q,\lambda)$. For direct excitons (neglecting polariton effects) they are obtained from Eq. (4.24) as

$$g_{K,\alpha}(q,\lambda) = ie_\lambda \cdot G_\alpha(q,\lambda) \delta_{K,q} . \tag{5.100}$$

As was discussed in Sect. 4.2.4, this relation can be extended to indirect excitons by the introduction of effective transition matrix elements, whereby the dynamics of the momentum conserving phonons need not to be considered. In this case one has to keep in mind that exciton states with different wavevector K do not interfere. This implies that all calculations have to be performed for a fixed, but arbitrary value of K and in the final step the scattered intensities have to be summed over all possible K vectors.

For arbitrary electronic systems, e.g. for disorder localized exciton states, the definition of the coupling factor can be generalized to a volume integral over the sample as

$$g_{K,\alpha}(\boldsymbol{k}, \lambda) = -i \int_{V_{sample}} d^3r \, E_\alpha \left(\frac{2\pi}{L^3 \omega_k} \right)^{1/2} \boldsymbol{M}_{K,\alpha} \cdot \boldsymbol{e}_\lambda e^{i\boldsymbol{k} \cdot \boldsymbol{r}} . \qquad (5.101)$$

Here the transition dipole moment is

$$\boldsymbol{M}_{K,\alpha} = < B_{K,\alpha}^\dagger \Phi_0 | D | \Phi_0 > \qquad (5.102)$$

with dipole operator D and ground state Φ_0. This allows one to calculate the coupling factors also for more complicated systems.

In order to apply the perturbation theory, it is of course necessary that the exciton coupling to the light field is weak implying that polariton effects have to be neglected. This restricts the possible applications to systems of localized exciton states as in quantum well structures and to indirect excitons. However, by applying the Hopfield transformation, the strong exciton photon interaction is renormalized and the perturbation theory can be applied directly to the polariton states interacting only with the thermal bath. This procedure is used in Chap. 7 to analyse the light scattering via exciton–polaritons in Cu_2O.

5.3.2 Calculation of the Time–Resolved Spectrum

Due to the equivalence of the classical equations with those of the Heisenberg picture of quantum mechanics, the relations derived in Chap. 2 immediately give the following quantum mechanical expression for the time–resolved spectrum

$$R(\omega_S, t) = \int_{-\infty}^{\infty} dt' \int_{-\infty}^{\infty} dt'' F_S^*(t - t') F_S(t - t'') \langle E_S^{(-)}(t') \, E_S^{(+)}(t'') \rangle \quad (5.103)$$

where the expression in brackets is the quantum mechanical field correlation function with $E_S^{(+)}(t)$ denoting the positive frequency part of the electric field operator of the scattered light.

Inserting the usual mode expansion of the field operator [106]

$$E^{(+)}(t) = \sum_{q,\lambda} i \left(\frac{\hbar \omega_q}{2\varepsilon_0 L^3} \right)^{1/2} \boldsymbol{e}_\lambda a_{q,\lambda}(t) e^{i\boldsymbol{q} \cdot \boldsymbol{r}} \qquad (5.104)$$

into the field correlation function and taking into account that by the spectral and spatial selection of the optical filter only modes with a certain wavevector q_S and polarization e_S are observed we obtain the following expression for the field correlation function

$$\langle E_S^{(-)}(t')\, E_S^{(+)}(t'') \rangle \propto \mathrm{Tr}[\rho^H a_{k_S,\lambda_S}^\dagger(t') a_{k_S,\lambda_S}(t'')]\ . \qquad (5.105)$$

Here ρ^H denotes the time independent statistical operator of the Heisenberg picture.

To proceed further in the calculation, we transform this equation into the interaction picture with undisturbed Hamiltonian H_0 whereby both pictures coincide at t_0. Performing the limit $t_0 \to -\infty$ we assume that no interaction exists at this time between the exciting laser pulse and the excitons; the density matrix of the total system can then be decomposed into the separate density matrices of the partial systems as

$$\rho(t_0 \to -\infty) = \rho_R^L \otimes \rho_{\mathrm{ex}}^0 \otimes \rho_B^{th}. \qquad (5.106)$$

Here ρ_B^{th} is the density matrix of the bath in thermal equilibrium, $\rho_{\mathrm{ex}}^0 = |\Phi_0\rangle\langle\Phi_0|$ the density matrix of the excitonic system in the ground state and ρ_R^L that of the exciting laser pulse.

The temporal dynamics is fully determined by the time–development operator $U(t, t_0)$ given by [147]

$$U(t, t_0) = \exp\left(-iH_0 \cdot (t - t_0)\right) T^{(+)} \exp\left(i \int_{t_0}^t V(t')dt'\right) \qquad (5.107)$$

with the Hamiltonian in the interaction picture

$$V(t) = e^{iH_0 t}\, V\, e^{-iH_0 t} \qquad (5.108)$$

and the positive time–ordering operator $T^{(+)}$. Using the cyclic commutativity of operators under the trace operation we obtain for the right side of Eq. (5.105)

$$\langle E_S^{(-)}(t')\, E_S^{(+)}(t'') \rangle = \langle k_S\lambda_S|\, \mathrm{Tr}_B[U(t', t_0)\rho(t_0)U^\dagger(t'', t_0)]\, |k_S\lambda_S\rangle \qquad (5.109)$$

that inserted into Eq. (5.103) gives the quantum mechanical expression for the *time–resolved spectrum*, i.e. the probability that at time t a photon will be detected in the light field under observation. In Eq. (5.109) the trace operation is to be performed only over the bath states.

With the two Eqs. (5.103) and (5.109) we have derived a very general expression for the time–resolved spectrum, which in contrast to other theoretical expressions [17, 144, 19, 10], shows in a clear way the influence of the detection process on the measurement, without making problematic assumptions, like an infinitely distant detector [19].

For the calculation of the time–dependent scattered intensity one has to construct from the stationary modes of the exciting light $a_{q\lambda}$ appropriate wavepackets to model the excitation laser pulse. They are assumed as a coherent superposition of photons with wavevector in the direction of the exciting laser beam ($q \parallel k_L$) with coefficients $f_{k_L'}$, that have a sharp maximum at $\omega_L = \omega_{k_L}$.

$$|k\lambda >= A^\dagger_{k\lambda}|0 > \tag{5.110}$$

with

$$A^\dagger_{k\lambda} = \sum_{k'\|k} f_{k'} a^\dagger_{k'\lambda} . \tag{5.111}$$

The coefficients are required to fulfil the relation

$$\sum_{k'\|k} |f_{k'}|^2 = 1 \tag{5.112}$$

in order that the creation and annihilation operators of the superposition photon states $A_{k\lambda}$ obey the Bose commutation relations. These wavepackets are specified by the time of excitation and their temporal shape. In the SVAP approximation this can be done by the field envelope functions as

$$F_L(t) = F_L^{(0)}(t)e^{-i\omega_L t} = \sum_{k'_L} f_{k'_L} e^{-i\omega_{k'_L} t} . \tag{5.113}$$

The density operator of these coherent states is then given by

$$\rho_R^L(t_0) = |k_L\lambda_L\rangle \langle k_L\lambda_L| . \tag{5.114}$$

5.3.3 Feynman Diagrams

The perturbational calculation of the expectation value in Eq. (5.109), where we use the procedure of Ref. [10], starts from the expansion of the time–development operator

$$U(t,t_0) = \sum_n \frac{(-i)^n}{n!} \int_{t_0}^t dt_1 \cdots \int_{t_0}^t dt_n T^{(+)}\{V(t_1) \cdots V(t_n)\} . \tag{5.115}$$

Inserting this expression into Eq. (5.109) leads to a sum over all possible interaction processes. The evaluation of these terms can be done using the Feynman diagram technique. Because of the twofold occurrence of U these diagrams represent *correlators* of two coupled time sequences corresponding to the positive and negative time–ordering operator $T^{(+)}$ and $T^{(-)}$. These rules were set up in the past by different authors (see e.g. [10, 148]) and are shown graphically in Fig. 5.7.

In the case of weak interaction between exciton and bath states, only the ladder diagrams shown in Fig. 5.8 contribute to the scattered intensity [10, 146, 148]. To simplify the calculations, the following additional assumptions have been made

- The memory time of the bath is short compared with the time–scales of the exciton and photon system (Markov approximation).

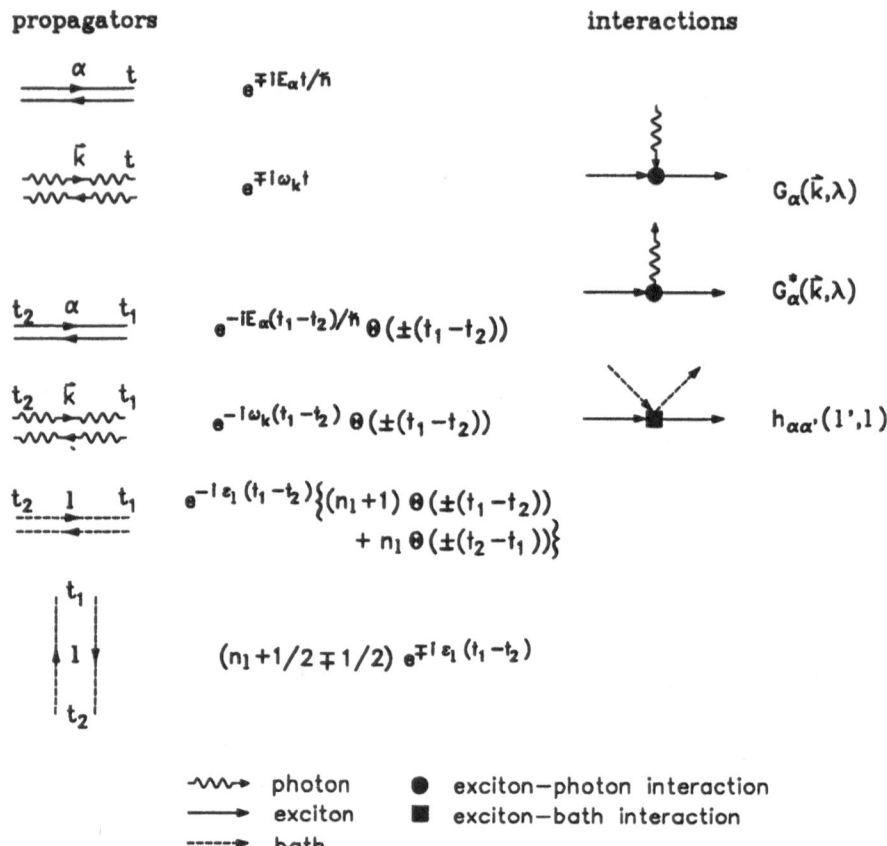

propagators   interactions

Fig. 5.7 Rules for the evaluation of Feynman diagrams to describe time–resolved light scattering. After [10]. α: exciton states; l, l': bath states; k: wavevector of the photon; n_l: Bose-Einstein distribution

- The exciton propagator contains all relaxation processes of the exciton states as a complex self–energy with homogeneous linewidth Γ_α. In the Markov approximation [10] Γ can be decomposed according to the relation

$$\Gamma_{K,\alpha} \equiv 2/\tau_{\text{coh}} = \Gamma_{K,\alpha}^{ER} + w_{K,\alpha} \tag{5.116}$$

$$= 2\pi \sum_{k,\lambda} |g_{K,\alpha}(k,\lambda)|^2 \delta(E_\alpha(K) - \omega_k)$$

$$+ \Gamma_{ac}^{ER} \tag{5.117}$$

$$+ 2\pi \sum_{l,l'} \sum_{\substack{K',\alpha' \\ E_{K,\alpha} = E_{K',\alpha'}}} |h_{K,\alpha,K',\alpha'}(l,l')|^2 \, n_{l'}(n_l + 1)\delta(\epsilon_l - \epsilon_{l'}).$$

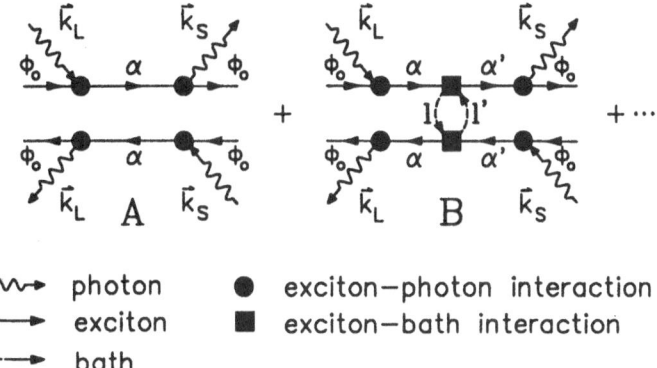

~~~→ photon   ● exciton–photon interaction

——→ exciton   ■ exciton–bath interaction

——→ bath

Fig. 5.8 Diagrammatic description of light scattering by ladder–type Feynman diagrams. Part A: coherent contribution; part B: incoherent scattering in lowest order (hot luminescence) Φ_0: ground state; α, α': exciton states; l, l': bath states; k_L, k_S: wavevector of the incident and scattered light.

in contributions to the energy relaxation Γ_α^{ER} comprising the radiative decay (first term) and relaxation processes by phonons (rate Γ_{ac}^{ER}, see Sect. 4.4), and in the pure phase relaxation by elastic scattering $w_{K,\alpha}$

$$w_{K,\alpha} = \sum_{K',\alpha'} \overline{h}_{K,\alpha,K',\alpha'} , \qquad (5.118)$$

where $\overline{h}_{K,\alpha,K',\alpha'}$ is the average matrix element of the exciton–bath interaction:

$$\overline{h}_{K,\alpha,K',\alpha'} = 2\pi \sum_{l,l'} |h_{K,\alpha,K',\alpha'}(l,l')|^2 \cdot n_{l'}(n_l + 1)\delta(\epsilon_l - \epsilon_{l'}) . \qquad (5.119)$$

Here n_l denotes the phonon occupation number. τ_{coh} is the coherence time of the excitons, which in this relaxation model coincides with the optical dephasing time T_2.

5.3.4 Coherent and Incoherent Scattering Contributions

For the zero–order diagram that contains no exciton–bath interaction vertex, the application of the rules in Fig.5.7 leads to the following expression

$$I^{(0)} = \sum_{K,\alpha,K',\alpha'} g_{K,\alpha}^*(k_S, e_S)g_{K,\alpha}(k_L, e_L)g_{K',\alpha'}(k_S, e_S)g_{K',\alpha'}^*(k_L, e_L)\cdot$$

$$\times \int_{-\infty}^{\infty} dt_1 \int_{-\infty}^{t_1} dt_2 \int_{-\infty}^{\infty} dt_1' \int_{-\infty}^{t_1'} dt_2' F_L(t_2)F_L^*(t_2')F_S^*(t - t_1)F_S(t - t_1')$$

$$\times e^{i\omega_S t_1} \cdot e^{-i\omega_S t_1'} \cdot e^{-i\omega_L t_2} \cdot e^{i\omega_L t_2'} \qquad (5.120)$$

$$\times e^{-i(E_{K,\alpha} - \frac{i}{2}\Gamma_\alpha)(t_1 - t_2)} \cdot e^{i(E_{K',\alpha'} - \frac{i}{2}\Gamma_{\alpha'})(t_1' - t_2')} .$$

Obviously, this diagram describes a scattering process proceeding without loss of coherence in the intermediate state. Therefore we denote $I^{(0)}$ in the following by I^{CS}.

A straightforward calculation gives the following expression for the case of direct excitons

$$I^{CS} = \left| \sum_{K,\alpha} g_\alpha^*(e_S) g_\alpha(e_L) \cdot \Psi_{0K,\alpha}(t) \right|^2 , \qquad (5.121)$$

where here the coherent process has to be identified with the Rayleigh scattering. In case of indirect excitons as intermediate states, where the scattering represents a Raman–type process, we obtain according to the discussion of the foregoing section

$$I^{CS} = \sum_K \left| \sum_\alpha g_\alpha^*(e_S) g_\alpha(e_L) \cdot \Psi_{0K,\alpha}(t) \right|^2 . \qquad (5.122)$$

In both cases the complex function $\Psi_{0K,\alpha}(t)$ is given by

$$\Psi_{0K,\alpha}(t) = \int_{-\infty}^{+\infty} dt_1 \int_{-\infty}^{+\infty} dt_2 \, F_L(t_2) F_S^*(t - t_1) \qquad (5.123)$$

$$\times e^{i\omega_S t_1 - i\omega_L t_2 - i(E_{K,\alpha} - \frac{i}{2}\Gamma_\alpha) \cdot (t_1 - t_2)} .$$

Comparing the expressions (5.121) and (5.122) shows that the light scattering of direct excitons is determined by the interference of all exciton states belonging to different K and therefore the time–dependence may deviate from that of the simple two–level system. For indirect excitons, on the other hand, the analogy can be quite close. However even here non–resonant contributions due to the K–summation also occur changing the overall time response.

The higher order ladder–diagrams represent those contributions to the scattered light that are emitted after the interaction of excitons with the bath states has occurred. They describe processes during which the phase of the excited states is lost. This part of the scattered intensity will therefore be named *hot luminescence*.

As an example of the general procedure the evaluation of the first order diagram will be presented in some detail. From the Feynman rules we obtain the expression

$$I^{(1)} = \sum_{K_1,\alpha_1,K_2,\alpha_2} |g_{K_1,\alpha_1}(k_S, e_S) g_{K_2,\alpha_2}(k_L, e_L)|^2$$

$$\times \int_{-\infty}^{\infty} dt_1 \int_{-\infty}^{t_1} dt_2 \int_{-\infty}^{t_2} dt_3 \int_{-\infty}^{\infty} dt_1' \int_{-\infty}^{t_1'} dt_2' \int_{-\infty}^{t_2'} dt_3'$$

$$\left\{ F_L(t_3) F_L^*(t_3') F_S^*(t - t_1) F_S(t - t_1') \cdot e^{i\omega_S(t_1 - t_1')} \cdot e^{-i\omega_L(t_3 - t_3')} \right.$$

$$\times e^{-i(E_{K_1,\alpha_1}-\frac{i}{2}\Gamma_{\alpha_1})(t_1-t_2)} \cdot e^{i(E_{K_1,\alpha_1}+\frac{i}{2}\Gamma_{\alpha_1})(t_1'-t_2')} \tag{5.124}$$

$$\times e^{-i(E_{K_2,\alpha_2}-\frac{i}{2}\Gamma_{\alpha_2})(t_2-t_3)} \cdot e^{i(E_{K_2,\alpha_2}+\frac{i}{2}\Gamma_{\alpha_2})(t_2'-t_3')}$$

$$\times \left(\sum_{l,l'} |h_{K_1,\alpha_1,K_2,\alpha_2}(l,l')|^2 \cdot e^{i((\epsilon_l-\epsilon_{l'})(t_2'-t_2)} \cdot n_{l'}(n_l+1) \right) \Bigg\} \ .$$

In the last round bracket the Markov approximation for the exciton–bath interaction will used in the form [10]

$$\sum_{l,l'} |h_{K_1,\alpha_1,K_2,\alpha_2}(l,l')|^2 \cdot n_{l'}(n_l+1) \cdot e^{i((\epsilon_l-\epsilon_{l'})(t_2'-t_2)} = \delta(t_2-t_2') \cdot \overline{h}_{K_1,\alpha_1,K_2\alpha_2}$$

$$\tag{5.125}$$

where $\overline{h}_{K,\alpha,K',\alpha'}$ is given by Eq. (5.119). This results in the following relation for the first order diagram

$$I^{(1)}(t) = \sum_{K_1,\alpha_1,K_2,\alpha_2} \overline{h}_{\alpha_1\alpha_2} |g_{\alpha_1}(e_S)g_{\alpha_2}(e_L)|^2$$

$$\times \int_{-\infty}^{+\infty} dt_2 e^{(\Gamma_{\alpha_1}-\Gamma_{\alpha_2})t_2} |\Psi_{1,\alpha_1}(t,t_2)|^2 \cdot |\Psi_{2,\alpha_2}(t_{n+1})|^2 \tag{5.126}$$

with the abbreviations

$$\Psi_{1,\alpha_1}(t,t') = \int_{t'}^{\infty} dt'' F_S^*(t-t'') \cdot e^{i\omega_S t'' - i(E_{K_1,\alpha_1}-\frac{i}{2}\Gamma_{\alpha_1})t''} \tag{5.127}$$

and

$$\Psi_{2,\alpha_2}(t') = \int_{-\infty}^{t'} dt'' F_L(t'') \cdot e^{-i\omega_L t'' + i(E_{K_2,\alpha_2}-\frac{i}{2}\Gamma_{\alpha_2})t''} \ . \tag{5.128}$$

In the same way the higher order diagrams lead to the n–th order intensity contribution

$$I^{(n)}(t) = \sum_{K_1,\alpha_1\cdots K_{n+1},\alpha_{n+1}} \overline{h}_{\alpha_1\alpha_2}\cdots\overline{h}_{\alpha_n\alpha_{n+1}} |g_{\alpha_1}(e_S)g_{\alpha_{n+1}}(e_L)|^2$$

$$\times \int_{-\infty}^{+\infty} dt_2 e^{(\Gamma_{\alpha_1}-\Gamma_{\alpha_2})t_2} |\Psi_{1,\alpha_1}(t,t_2)|^2 \tag{5.129}$$

$$\times \int_{-\infty}^{t_2} dt_3 e^{(\Gamma_{\alpha_2}-\Gamma_{\alpha_3})t_3} \int_{-\infty}^{t_3} dt_4 \cdots$$

$$\cdots \int_{-\infty}^{t_n} dt_{n+1} e^{(\Gamma_{\alpha_n}-\Gamma_{\alpha_{n+1}})t_{n+1}} |\Psi_{2,\alpha_{n+1}}(t_{n+1})|^2 \ .$$

Summation over all orders gives as total intensity of the *hot luminescence*

$$I^{HL} = \sum_{n=1}^{\infty} I^n(t) \ . \tag{5.130}$$

From this derivation we see that the bath interaction causes the phase of the intermediate states to be lost (this occurs instantaneously in the Markov approximation) so that the hot luminescence is an incoherent process. The difference between the coherent and incoherent processes is visible clearly in the way of summation in the two expressions (5.121) and (5.129). While in the first process first the summation and then the squaring is performed and the relative phase of the intermediate states is of importance, for the hot luminescence processes we have to first take the absolute square, whereby any phase information is lost, and then sum as the final step. From this we conclude that the Rayleigh and the Raman contribution to the scattering process reflects the coherent nature of the exciton states.

As can be seen from Eqs. (5.121) and (5.123), the temporal dependence of the coherent part is determined by the coherence lifetime τ_{coh} of the states allowing a direct measurement of the homogeneous linewidth of the states. The hot luminescence contribution depends via $\bar{h}_{K,\alpha,K',\alpha'}$ in addition on the pure phase relaxation rate w (Eq. (5.118)). Comparing the temporal behaviour of both parts one can obtain with Eq. (5.116) also the energy relaxation rate Γ^{ER}. In this way, all relaxation parameters of an exciton system can be obtained from one experimental method. This is a distinct advantage of time–resolved light scattering compared to other methods like four–wave–mixing where only the coherent relaxation rates are easily measurable.

To discriminate between the coherent and incoherent contributions to the scattered light, the dependence of the intensity on light polarization given by the coupling factors $g_{K_1,\alpha_1}(k_S, e_S)$ can be used with advantage, as this turns out to be different for the coherent and incoherent contributions. Moreover, this effect does not depend on the detailed shape of the excitation pulse shape or on the detection setup, but only on the symmetry properties of the exciton states.

In this way, light scattering enables one to monitor the temporal evolution from the initially excited coherent state into an incoherent ensemble of excitons, the information about the coherence being contained in the Rayleigh and Raman component. This shows up very clearly if we consider the resonant excitation of an ensemble of energetically split states. Then it follows immediately from Eq. (5.121) that the coherent part of the scattered intensity oscillates with frequencies given by the energy differences of the states. Obviously the hot luminescence does not show such oscillations indicating that the coherence is lost by phase relaxation. Therefore *quantum beats* occur only in the Rayleigh or Raman scattered light. This allows one to *experimentally* discriminate between these processes, since it is well known that Raman scattering in the proper sense comprises that part of the scattered light emitted with conservation of the phase in the intermediate state, while hot luminescence contains all other processes where the phase is lost.

The discrimination between the coherent and incoherent scattering processes by the observation of quantum beats will be demonstrated for different systems (see Chaps. 6 and 7) and fully confirms the theoretical expectations.

In contrast to this method, the discrimination of Raman and hot luminescence processes simply by their temporal behaviour is much more difficult as it needs a complicated fitting procedure of all components of the emitted light. The simple assumption that Raman (or Rayleigh) processes proceed instantaneously, which is usually made (see e.g. [149]), is only a very rough approximation and in some cases may be completely wrong.

5.3.5 Model Calculations

In this section the relations derived above will be applied to some simple cases that allow a complete analytic solution. These are then used to show how the results of the transform–limited quantum theory differ from those obtained by the density matrix treatment of the last section, which is used in the form of rate equations in the analysis of most of the experimental observations of resonant light scattering from excitons in semiconductors.

To keep the discussion as simple as possible, we will assume that the total relaxation rates of all exciton states are equal $\Gamma_\alpha = \Gamma_\alpha^{ER} + w_\alpha = \Gamma = const$, and that the elastic scattering rates $\bar{h}_{\alpha_1,\alpha_2}$ are also constant for all states.

In the first step we discuss only one K state. This is already a useful approximation if due to selection rules the exciton–photon interaction is restricted to one state, as is the case for localized excitons. For all other cases one has to sum the results over a certain range of wavevectors K, like for Raman scattering by indirect excitons. Thereby one must observe the important rule that the summation has to be performed incoherently if there is the possibility to discriminate between the different channels after the scattering process *in principle*.

Under these assumptions the hot luminescence component Eq. (5.130) can be written as

$$I^{HL} = \frac{1}{N} \sum_{\alpha_1,\alpha_2} |g_{\alpha_1}(e_S)g_{\alpha_2}(e_L)|^2 \, A^{HL}_{\alpha_1,\alpha_2,K}(t) \tag{5.131}$$

with

$$A^{HL}_{\alpha_1,\alpha_2,K}(t) = \int_{-\infty}^{+\infty} dt_2 |\Psi_1(t,t_2)|^2 \tag{5.132}$$

$$\times \left[w|\Psi_2(t_2)|^2 + \sum_{n=2}^{\infty} w^n \int_{-\infty}^{t_2} dt_3 \cdots \int_{-\infty}^{t_n} dt_{n+1} |\Psi_2(t_{n+1}|^2 \right]$$

where N denotes the number of excited exciton states.

In the simplest case both the excitation pulse and the response function of the detector have a delta–function like shape

$$F_L(t) = \delta(t) \tag{5.133}$$

$$F_S(t) = \delta(t) . \tag{5.134}$$

In this case the time–resolved spectrum does not contain any spectral information ('white–light limit'). Furthermore the exciton states are all assumed to be degenerate $E_{K,\alpha} = E_0 - i/2\Gamma$. Then inserting the Eqs. (5.133) and (5.134) into the relation (5.121) gives for the Rayleigh (or Raman) intensity

$$I^{RS} = \left| \sum_\alpha g_\alpha^*(e_S) g_\alpha(e_L) \right|^2 \cdot e^{-\Gamma t} . \tag{5.135}$$

For the hot luminescence one obtains

$$I^{HL} = \frac{1}{N} \sum_{\alpha_1,\alpha_2} |g_{\alpha_1}(e_S) g_{\alpha_2}(e_L)|^2 \cdot \left(e^{-(\Gamma-w)t} - e^{-\Gamma t} \right) . \tag{5.136}$$

To specialize these relations to the standard model system of two degenerate excited states ($N = 2$) with orthogonal transition moments (see Sect. 5.2.3), the excitation polarization e_L is chosen so that

$$e_L \cdot \hat{M}_1 = 1 \qquad e_L \cdot \hat{M}_2 = 0 . \tag{5.137}$$

The polarization directions of the scattered light $e_S^\parallel$ and $e_S^\perp$ are defined by

$$e_S^\parallel = e_L \qquad e_S^{\perp*} \cdot e_L = 0 . \tag{5.138}$$

Then we obtain for the Raman intensity

$$I^{RS}(e_S^\parallel) = e^{-\Gamma_\alpha t} \tag{5.139a}$$
$$I^{RS}(e_S^\perp) = 0 \tag{5.139b}$$

while the hot luminescence intensity is

$$I^{HL}(e_S^\parallel) = \frac{1}{2} \left(e^{-(\Gamma_\alpha-w)t} - e^{-\Gamma_\alpha t} \right) \tag{5.140a}$$
$$I^{HL}(e_S^\perp) = \frac{1}{2} \left(e^{-(\Gamma_\alpha-w)t} - e^{-\Gamma_\alpha t} \right) . \tag{5.140b}$$

The total intensity of the scattered light follows as

$$I(e_S^\parallel) = \frac{1}{2} \left(e^{-(\Gamma_\alpha-w)t} + e^{-\Gamma_\alpha t} \right) \tag{5.141a}$$
$$I(e_S^\perp) = \frac{1}{2} \left(e^{-(\Gamma_\alpha-w)t} - e^{-\Gamma_\alpha t} \right) . \tag{5.141b}$$

These are the same relations as already derived in the phenomenological density matrix formalism which is obtained here as the limiting case of the transform–limited theory for broad–band excitation and detection, the *white–light limit*. Obviously, in this limit, the conceptual difference between the coherent scattering process and a picture of resonant light scattering as *absorption followed by emission* (see e.g. [49, 51]) is no longer present and the distinction between Raman scattering and hot luminescence is not possible. It is the assumption of the white–light limit, which is implicit in all these calculations, that finally forces the independence of the absorption from the emission process.

More realistic and useful for a comparison with experimental results is to choose a real transform–limited excitation pulse and detection setup. Here we assume the following envelope function for the excitation

$$F_L(t) = \sqrt{\gamma_L} \cdot \Theta(t) \cdot e^{-\gamma_L t/2}, \qquad (5.142)$$

while the detection response function is given by

$$F_S(t) = \sqrt{\gamma_S} \cdot \Theta(t) \cdot e^{-\gamma_S t/2} . \qquad (5.143)$$

These relations describe a single–sided laser pulse and a Fabry–Perot resonator as spectral filter (see Sect. 2.1.3). $\Theta(t)$ denotes the Heavyside step function. The quantities γ_L and γ_S enable the temporal and spectral behaviour of laser pulse and detection to be varied.

The coherent part (zero–order diagram) is completely specified by the function $\Psi_{0,\alpha}(t)$, which represents the exciton propagator. This is obtained as

$$\Psi_{0,\alpha}(t) = \frac{e^{-\frac{\gamma_S}{2}t}}{z_{lk}} \cdot \left[\frac{e^{z_{sk}t}}{z_{sk}} + \frac{e^{-z_{ls}t}}{z_{ls}} - \left(\frac{1}{z_{sk}} + \frac{1}{z_{ls}} \right) \right] \qquad (5.144)$$

with the abbreviations

$$
\begin{aligned}
z_{lk} &= i(\omega_L - E_\alpha) + \frac{1}{2}(\gamma_L - \Gamma_\alpha) \\
z_{sk} &= i(\omega_S - E_\alpha) + \frac{1}{2}(\gamma_S - \Gamma_\alpha) \qquad (5.145) \\
z_{ls} &= i(\omega_L - \omega_S) + \frac{1}{2}(\gamma_L - \gamma_S) .
\end{aligned}
$$

The total coherently scattered intensity follows by summing over all exciton states and squaring this result.

For the contribution due to hot luminescence the following expression can be derived after a somewhat lengthy and tedious calculation

$$I^{HL} = \frac{1}{N} \sum_{\alpha_1, \alpha_2} |g_{\alpha_1}(e_S) g_{\alpha_2}(e_L)|^2 \, \gamma_L \cdot \gamma_S \cdot A_{lk,\alpha_1} \cdot A_{sk,\alpha_2} \cdot A_{\alpha_1,\alpha_2}(t) . \qquad (5.146)$$

Here the following abbreviations have been introduced

$$A_{lk,\alpha_1} = \frac{1}{(\omega_L - E_{\alpha_1})^2 + \frac{1}{4}(\gamma_L - \Gamma)^2} \qquad (5.147)$$

$$A_{sk,\alpha_2} = \frac{1}{(\omega_S - E_{\alpha_2})^2 + \frac{1}{4}(\gamma_S - \Gamma)^2} \qquad (5.148)$$

and

$$
\begin{aligned}
A_{\alpha_1,\alpha_2}(t) &= \left\{ w - w^2 \cdot \left[\frac{1}{\Gamma^{ER} - \gamma_L} + 2\Re e\left(\frac{1}{z_{lk} + w} \right) \right] \right\} \\
&\times \left[\frac{1}{w} \left(e^{-\Gamma^{ER}t} - e^{-\Gamma t} \right) + \frac{1}{\gamma_S - \Gamma^{ER}} \left(e^{-\Gamma^{ER}t} - e^{-\gamma_S t} \right) \right. \\
&\qquad \left. - 2\Re e\left(\frac{1}{z_{skp}^* - \Gamma^{ER}} \left(e^{-\Gamma^{ER}t} - e^{-z_{skp}^* t} \right) \right) \right] \\
&+ \frac{w \cdot (\Gamma - \gamma_L)}{\Gamma^{ER} - \gamma_L} \cdot \left[\frac{1}{\Gamma - \gamma_L} \left(e^{-\gamma_L t} - e^{-\Gamma t} \right) \right. \\
&\qquad \left. + \frac{1}{\gamma_S - \gamma_L} \left(e^{-\gamma_L t} - e^{-\gamma_S t} \right) - 2\Re e\left(\frac{1}{z_{skp}^* - \gamma_L} \left(e^{-\gamma_L t} - e^{-z_{skp}^* t} \right) \right) \right] \\
&- 2\Re e\left\{ \frac{z_{lk} \cdot w}{z_{lk} + w} \left[\frac{1}{z_{lk}} \left(e^{-\Gamma t} - e^{-z_{lkp} t} \right) + \frac{1}{\gamma_s - z_{lkp}} \left(e^{-z_{lkp} t} - e^{-\gamma_S t} \right) \right. \right. \\
&\qquad \left. \left. - \frac{1}{z_{skp}^* - z_{lkp}} \left(e^{-z_{lkp} t} - e^{-z_{skp}^* t} \right) - \frac{1}{z_{skp} - z_{lkp}} \left(e^{-z_{lkp} t} - e^{-z_{skp} t} \right) \right] \right\}
\end{aligned}
\tag{5.149}
$$

with

$$
z_{lkp} = i(\omega_L - E_{\alpha_1}) + \frac{1}{2}(\gamma_L + \Gamma), \quad z_{skp} = i(\omega_S - E_{\alpha_2}) + \frac{1}{2}(\gamma_S + \Gamma). \tag{5.150}
$$

In these relations N denotes the number of excited degenerate exciton states. However, in the case of an exciton symmetry (given by the representation Γ) where the product representation $\Gamma^* \otimes \Gamma$ contains each irreducible representation of the group at most once, the relaxation matrix has to fulfil certain symmetry conditions. For the components $\bar{h}_{\alpha_1,\alpha_2}$ of the relaxation matrix these have the consequences that the factor $1/N$ has to be omitted and that the pure phase relaxation rate w is given by the average of the $\bar{h}_{\alpha_1,\alpha_2}$ and not by the sum as was assumed above. Of all systems investigated here, this only applies to the $1S$–exciton in Cu_2O.

Finally, we will compare the results of the transform–limited theory for the coherent and incoherent contribution obtained for an excitonic system with that of the simple density matrix theory that was shown to be equivalent to the rate equation treatment. This is important to see whether the transform–limited treatment is different from the procedure common in the literature to obtain from measured decay curves information about the relaxation dynamics. As laser pulse we have chosen a single sided exponential with a decay constant γ_L and as spectral filter a Fabry–Perot resonator with spectral width γ_S. An additional time–broadening of the photodetector was assumed to be of gaussian form with a FWHM of 20 ps, closely modeling the actual setup (see Chap. 3).

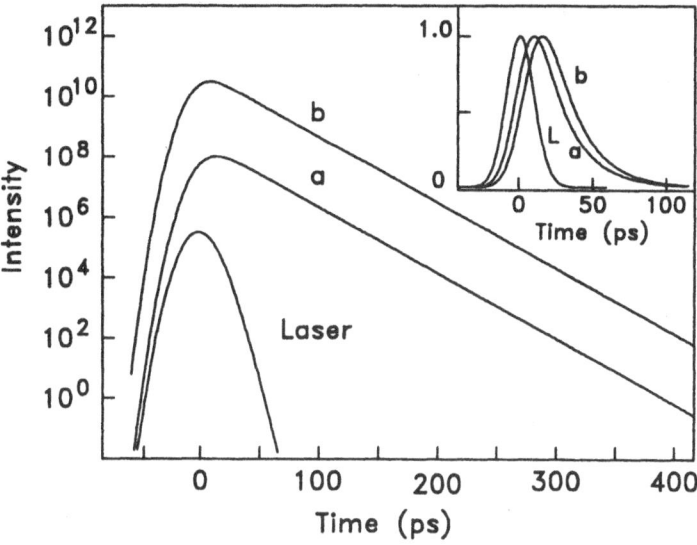

Fig. 5.9 Comparison of the results of the transform–limited theory with the standard procedure of obtaining time–resolved decay curves of resonance fluorescence. Curve L: response of the setup to the laser pulse; a: rate-equation response; b: transform–limited response. The intensity scale for each curve is shifted by a factor of 100. Further discussion see text.

In Fig. 5.9 the temporal behaviour is plotted for the rate equation result (curve a) and for the case of a single exciton state (curve b). The curve marked L is the time response of the setup to the excitation laser pulses. In the rate equation theory, the effect of the finite response time of the setup is taken into account by a convolution of the response of the system to the excitation laser pulse with the model function. This leads to curve b in Fig. 5.9. In the transform–limited result only the additional time–broadening of the detector needs to be considered according to Eq. (2.13) (curve a).

The relaxation rates are chosen in such a way that the pure dephasing amounts to 50% of the homogeneous linewidth ($\Gamma = 50 \cdot 10^9$ s^{-1}) and the excitation and detection is done with $\Gamma_L = 300$ GHz and $\Gamma_S = 350$ GHz corresponding to a time resolution of about 10% of the coherence time.

As expected the polarized component decays with the same apparent time constant. However, a significant difference is obtained for the onset of the scattered intensity, where the transform–limited theory predicts a delay in the rise of the signal (see inset of Fig. 5.9) This effect has been actually observed in the experiments reported in Chap. 8. Therefore one has to be very careful in interpreting time–resolved decay data simply by applying rate equation models.

6. Exciton Dynamics in Silver Bromide

6.1 Introductory Remarks

In recent years the silver halides, in particular silver bromide, have become model systems for the investigation of light scattering from exciton states allowing several phenomena to be observed for the first time (for reviews see [22, 42]). One reason for this is the fact that they possess an indirect gap in the blue spectral region that is accessible to dye laser systems. Furthermore, due to the indirect character of the exciton states with a dispersion minimum at the L point of the Brillouin zone (BZ), resonant light scattering processes involving these excitons are quite simple. In an ideal crystal, the scattering always involves two momentum conserving phonons coming into resonance at low temperatures for excitation energies above the indirect absorption edge.

In this material the first direct measurements of exciton relaxation by time–resolved spectroscopy were performed [44, 45] (for a review see [22]). These studies showed that it is possible to obtain the rates of the various exciton relaxation processes from such experiments and to derive the corresponding electron–phonon interaction parameters with high precision.

In these investigations, pronounced polarization effects were observed in the intensities of the two–phonon scattering processes. These had already been found in stationary measurements [43], but could not be understood at that time. Here the time–resolved investigations have shown that the scattered intensities always consist of a polarized and an unpolarized contribution. As the depolarization times were found to be surprisingly long, of the order of several hundreds of picoseconds, these investigations suggested the possibility of observing *quantum beats* from these exciton states in the time–dependent light scattering. Actually, the indirect exciton in AgBr was the first system where quantum beats from an extended electronic state in a solid were found [150, 151].

In this chapter the basic notions of resonant light scattering processes in silver halides will be treated in a very systematic way to serve as a model for further studies of this kind, which can be done in other systems like silicon, germanium and other indirect gap materials. This requires the following steps

- determination of the exciton states and their energies in external fields by a group–theoretical analysis applying the method of invariants,

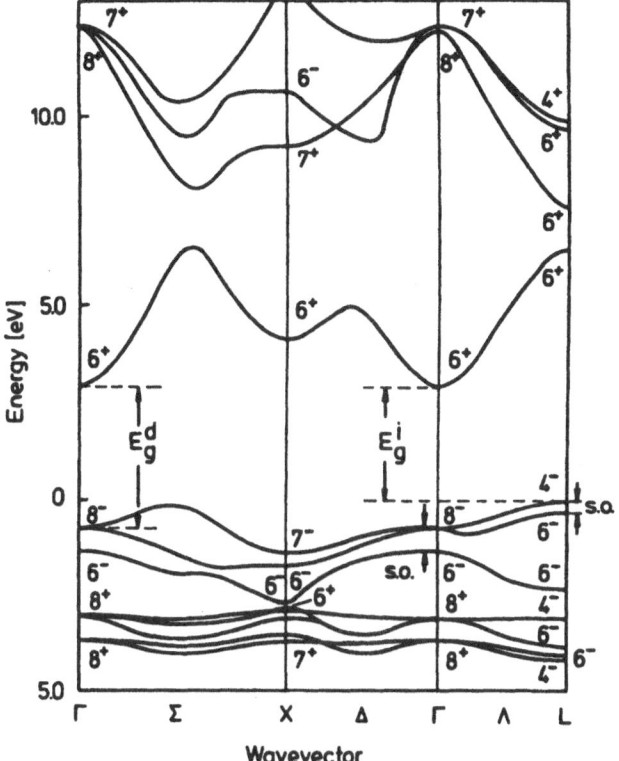

Fig. 6.1 Band structure of AgBr after [154]. Origin of coordinate system at the Cl⁻ ion.

- calculation of the optical transition moments of the indirect exciton states,
- comparison of model calculations of the scattered intensities using the methods developed in Chap. 5 with the experimental results.

6.2 Exciton States

6.2.1 Band Structure

The starting point for the following discussion of the exciton states is of course the band structure of the silver halides AgBr and AgCl both crystallizing in the sodium chloride structure with space group O_h^5 and point group O_h. In the past, several band structure calculations have been performed using e.g. a semiquantitative 'tight-binding' model [152], the APW method ([153]), and, more recently, KKR calculations [154]. The result of the KKR calculations, which include the spin–orbit coupling, is shown in Fig. 6.1. According to these calculations, the minimum of the conduction band is in the centre of the BZ and has, in agreement with the participation of Ag^+5s wavefunctions Γ_6^+ symmetry [152]. The symmetry–adapted states can therefore be taken in the form

$$\psi_{1/2}^6 = S_c\alpha_c \qquad \text{and} \qquad \psi_{-1/2}^6 = S_c\beta_c \ . \qquad (6.1)$$

Here α_c and β_c denote the spin functions with $s_z = 1/2$ and $s_z = -1/2$. S_c represents a spatial function that transforms according to the representation Γ_1 of the symmetry group O_h. The same conduction band transforms at the L point of the BZ according to the representation $L_6^-(L_2^-)$ of the symmetry group D_{3d} (the notation in brackets is without spin). Conduction bands higher in energy that are important as virtual states for the indirect transition are those of symmetry $\Gamma_8^+ + \Gamma_7^+(\Gamma_5^+)$ at the Γ-point, which split at the L point into the states $L_6^+ + L_{4,5}^+(L_3^+)$ and $L_6^+(L_1^+)$.

The highest valence band states are combinations of halogen p and silver d states [152]. As this mixing is forbidden by symmetry at the Γ point, here the highest valence bands have pure halogen p character. Without spin–orbit coupling their symmetry is therefore Γ_4^-, leading to $\Gamma_8^- + \Gamma_6^-$ states taking the spin into account. These valence band states therefore correspond to angular momentum eigenstates with $j = 1/2$ (Γ_6^-) and $j = 3/2$ (Γ_8^-) given by[1]

$$\left.\begin{aligned}
\phi_{1/2}^6 &= -\frac{i}{\sqrt{3}}[Z_\Gamma\alpha_v + (X_\Gamma + iY_\Gamma)\beta_v] \\
\phi_{-1/2}^6 &= -\frac{i}{\sqrt{3}}[Z_\Gamma\beta_v - (X_\Gamma - iY_\Gamma)\alpha_v]
\end{aligned}\right\} \qquad \text{for} \quad \Gamma_6^- \qquad (6.2)$$

and

$$\left.\begin{aligned}
\phi_{3/2}^8 &= -\frac{i}{\sqrt{2}}(X_\Gamma + iY_\Gamma)\alpha_v \\
\phi_{1/2}^8 &= \frac{i}{\sqrt{3}}\left[\sqrt{2}Z_\Gamma\alpha_v - \frac{1}{\sqrt{2}}(X_\Gamma + iY_\Gamma)\beta_v\right] \\
\phi_{-1/2}^8 &= \frac{i}{\sqrt{3}}\left[\sqrt{2}Z_\Gamma\beta_v + \frac{1}{\sqrt{2}}(X_\Gamma - iY_\Gamma)\alpha_v\right] \\
\phi_{-3/2}^8 &= \frac{i}{\sqrt{2}}(X_\Gamma - iY_\Gamma)\alpha_v
\end{aligned}\right\} \qquad \text{for} \quad \Gamma_8^- \ . \qquad (6.3)$$

Here the spatial functions X_Γ, Y_Γ and Z_Γ are a basis of the representation Γ_4^- of the group O_h transforming like the corresponding coordinates. These are assumed in the following to be parallel to the cubic axes of the fcc structure [100], [010] and [010], but due to the still present rotational symmetry of the states any other choice is possible. The Γ_8^- and Γ_6^- states are split by the spin–orbit coupling, which is of the same magnitude as in the free halogen

---

[1]To denote valence band states we will use in the following the so called 'missing–electron' scheme, where the state of the electron taken out from the valence band is specified.

Table 6.1 Coordinate systems of the four inequivalent L points of the fcc lattice.

L point	axis	direction	L point	axis	direction
	x_1	$\frac{1}{\sqrt{6}}(-1,-1,2)$		x_2	$\frac{1}{\sqrt{6}}(1,-1,2)$
1	y_1	$\frac{1}{\sqrt{2}}(1,-1,0)$	2	y_2	$\frac{1}{\sqrt{2}}(1,1,0)$
	z_1	$\frac{1}{\sqrt{3}}(1,1,1)$		z_2	$\frac{1}{\sqrt{3}}(-1,1,1)$
	x_3	$\frac{1}{\sqrt{6}}(1,1,2)$		x_4	$\frac{1}{\sqrt{6}}(-1,1,2)$
3	y_3	$\frac{1}{\sqrt{2}}(-1,1,0)$	4	y_4	$\frac{1}{\sqrt{2}}(-1,-1,0)$
	z_3	$\frac{1}{\sqrt{3}}(1,1,-1)$		z_4	$\frac{1}{\sqrt{3}}(1,-1,1)$

atom. From absorption and reflection studies it is known to be about 0.16 eV in AgCl and 0.64 eV in AgBr [155].

Outside the Γ point the silver d states also contribute to the uppermost valence band according to the relations $D_1^-(X^-) \xrightarrow{L} L_2^- + L_3^-$ and $D_2^+(Ag^+) \xrightarrow{L} L_2^- + 2L_3^-$) (with D_1^- and D_2^+ denoting the representations of the rotational group for $l = 1,2$)[2]. In the highest valence band anti–bonding silver d states are mixed with the halogen states thus pushing the maximum of the band to the L point of the BZ.

Neglecting the spin–orbit interaction one has the following compatibility relations

$$\Gamma_4^- \xrightarrow{\Lambda} \Lambda_1 + \Lambda_3 \xrightarrow{L} L_2^- + L_3^- , \tag{6.4}$$

which lead to Bloch states at the L point of the form

$$\left. \begin{array}{rcl} \phi_+^3 &=& -\frac{i}{\sqrt{2}}(X_L + iY_L) \\[2mm] \phi_-^3 &=& +\frac{i}{\sqrt{2}}(X_L - iY_L) \end{array} \right\} \qquad \text{for} \quad L_3^- \tag{6.5}$$

$$\phi^2 = iZ_L \qquad\qquad\qquad L_2^- . \tag{6.6}$$

Here the functions X_L, Y_L, Z_L transform like the coordinate axis of the L point under consideration. As there are four inequivalent L points in the BZ forming the star of $\mathbf{k}_L$, the choice of the coordinate axis is partly determined by symmetry. Here we will always refer to the coordinate systems given in Table 6.1.

---

[2] The notation depends here also on the choice of the origin [153]. Here we have chosen in agreement with [152] the origin at the Cl^- ion .

Including the spin–orbit interaction one has the relations

$$\Gamma_8^- \xrightarrow{\Lambda} \Lambda_4 + \Lambda_5 + \Lambda_6 \xrightarrow{L} L_4^- + L_5^- + L_6^- \tag{6.7}$$

and

$$\Gamma_6^- \xrightarrow{\Lambda} \Lambda_6 \xrightarrow{L} L_6^- . \tag{6.8}$$

Here unlike Koster [100], we denote the one–dimensional Kramer's degenerate spinor representations by $L_{4,5}$ and the two–dimensional representation as L_6. The corresponding Bloch states are given according to Eq. (6.5) as

$$\left.\begin{aligned}
\phi^4 &= -\frac{i}{\sqrt{2}}(X_L + iY_L)\alpha_v \\
\phi^5 &= \frac{i}{\sqrt{2}}(X_L - iY_L)\beta_v
\end{aligned}\right\} \qquad L_{4,5}^- \tag{6.9}$$

$$\left.\begin{aligned}
\phi_+^6 &= -\frac{i}{\sqrt{2}}(X_L + iY_L)\beta_v \\
\phi_-^6 &= \frac{i}{\sqrt{2}}(X_L - iY_L)\alpha_v
\end{aligned}\right\} \qquad L_6^- \tag{6.10}$$

for the states compatible with the Γ_8^- states, while the Γ_6^- compatible states are given by

$$\left.\begin{aligned}
\phi_+^{6'} &= iZ_L\alpha_v \\
\phi_-^{6'} &= iZ_L\beta_v
\end{aligned}\right\} \qquad L_6^- \tag{6.11}$$

Comparing these results with relations (6.3) shows that the $L_{4,5}$–states are pure angular momentum states with $m_j = \pm 3/2$ while the L_6 states are mixtures of angular momentum functions for $j = 3/2$ and $j = 1/2$ with $m_j = \pm 1/2$. It should be noted that the relations (6.10) and (6.11) are exactly valid only in case of vanishing spin–orbit splitting. As the spin–orbit induced mixing does not occur for the most important $L_{4,5}^+$ states, its effect need not to be discussed further. Experimentally, the splitting between the $L_{4,5}^-$ and L_6^- states is not known very precisely. From absorption measurements in AgBr and AgCl [116, 156] the following approximate values can be deduced; they are compared with those from the band structure calculation [154] (values given in brackets):

$$\text{AgCl}: E_{L_{4,5}^-} - E_{L_6^-} = 0.03 \text{ eV}(0.02 \text{ eV}) \tag{6.12}$$

$$\text{AgBr}: E_{L_{4,5}^-} - E_{L_6^-} = 0.25 \text{ eV}(0.28 \text{ eV}) \tag{6.13}$$

6.2.2 Indirect Excitons

The following discussion will be restricted to those exciton states that are obtained by coupling the lowest conduction band states given by relation (6.1) with the hole states derived from the L_3^- band (Eq. (6.5)). They transform according to the product representation $\Gamma_1^+ \otimes L_3^{-*}$. In the notation of the exciton states the missing–electron scheme is applied. In this scheme $\phi \otimes \psi$ denotes the Slater determinant for which the valence band state ϕ is empty and the conduction band state ψ is occupied, i.e. the state $\phi \otimes \psi = a_\psi^\dagger a_\phi |0\rangle$. In the calculations one should note that matrix elements between valence band states show up as the negative complex conjugate in the exciton problem [157]. In particular this implies that the hole angular momentum is equal to the negative momentum of the missing electron. For example, in the state denoted by ϕ^4 the hole angular momentum is $m_j^h = -3/2$.

The energies of the exciton states can be obtained from an effective Hamiltonian that depends on internal variables and external parameters like the magnetic field. The form of this operator is fully determined by the symmetry of the states and can be obtained most straightforwardly by the method of invariants (see e.g. [94, 157]). Thereby the terms in the Hamiltonian that depend on the electron and hole spin are also treated as an external perturbation like the wavevector k, magnetic field B or electric field E. All external operators are combined in a set Ξ.

The effective Hamiltonian of the subspace that is spanned by the functions (6.1) and (6.5) with dimension $n = 2$ can be expanded in a complete set of independent linear operators. Here we choose for this basis the set $\{\mathcal{I}_c\} \otimes \{\mathcal{I}_v, L_z, L_+, L_- = (L_z - iL_y)/2\}$. In matrix form these are given by the well–known Pauli matrices

$$\mathcal{I}_v = \begin{bmatrix} 1 & 0 \\ 0 & 1 \end{bmatrix} \quad L_z = \begin{bmatrix} 1 & 0 \\ 0 & -1 \end{bmatrix} \quad L_+ = \begin{bmatrix} 0 & 1 \\ 0 & 0 \end{bmatrix} \quad L_- = \begin{bmatrix} 0 & 0 \\ 1 & 0 \end{bmatrix} \quad (6.14)$$

and by

$$\mathcal{I}_c = [1] . \quad (6.15)$$

Here the operators $L_\pm$ are defined as the following linear combinations of the x and y operators

$$L_+ = (L_x + iL_y)/2 \quad L_- = (L_x - iL_y)/2 . \quad (6.16)$$

The terms depending on the electron and hole spin, s and σ, respectively, are expressed with the corresponding Pauli matrices $\{\mathcal{I}_s, s_z, s_+, s_-\}$ and $\{\mathcal{I}_\sigma, \sigma_z, \sigma_+, \sigma_-\}$.

All basis operators can be classified according to their transformation under the operations of the symmetry group and under time reversal. For the hole operators these are given in Table 6.2. The electron spin transforms according to the representation Γ_4^+ of the group O_h. The time reversal symmetry of the operators can be deduced from that of the basis functions themselves. From this follows for s, σ and L_z the type $\hat{K}_-$ (inversion under time reversal), while $L_\pm$ transforms with type $\hat{K}_+$ (even behaviour under time reversal).

Table 6.2 Classification of physical observables according to their transformation properties in the group D_{3d}.

$\mathcal{D}$		$\hat{K}_-$		$\hat{K}_+$				
Γ_1^+	S				B_z^2	$B_x^2+B_y^2$	k_z^2	$k_x^2+k_y^2$
Γ_2^+	T	B_z	σ_z	L_z				
Γ_3^+	U_+	B_+	σ_+	L_+				
	U_-	B_-	σ_-	L_-				
Γ_1^-	V							
Γ_2^-	W	k_z		E_z				
Γ_3^-	X_+	k_+		E_+				
	X_-	k_-		E_-				

Now the effective Hamiltonian is composed of all linear combinations of the basis operators (6.10) and (6.11) with the operators in Ξ that transform according to the fully symmetric representation Γ_1^+ of the symmetry group. Thereby the Hamiltonian $\mathcal{H}$ can be divided into three parts $\mathcal{H}^{(c)}+\mathcal{H}^{(v)}+\mathcal{H}^{(ex)}$. $\mathcal{H}^{(c)}$ contains linear combinations of the electron spin s and Ξ representing the conduction band energy. In the same way $\mathcal{H}^{(v)}$ describes the valence band energy and is composed of linear combinations of $\sigma,\{L_z,L_+,L_-\},\Xi$. All invariant linear combinations of $s,\sigma,\{L_z,L_+,L_-\}$, and Ξ comprise the excitonic exchange interaction $\mathcal{H}^{(ex)}$ between electron and hole.

The transformation properties of any operator in the symmetry group D_{3d} of the L point can be classified using the irreducible representations of this group. For this purpose 'dummy' operators $S,\ldots,X_\pm$ are used that serve as basis for the irreducible representations of D_{3d} (see Table 6.3) and have to be replaced by the physical observables in a concrete calculation. In this table the transformation properties of the operators L, of the wavevector k, of the electric and magnetic fields E and B and of the hole spin σ are given explicitly. While the electric field E and the wavevector k are polar vectors, the magnetic field B transforms as a pseudo-vector. In order to have a consistent description we use in this table for all vector operators O the representation with z component O_z and $O_\pm = \frac{1}{\sqrt{2}}(O_x \pm iO_y)$ the standard form of symmetry adapted linear combinations of x and y components [94].

The symmetry adapted linear combinations of these operators that are needed in the theory of invariants can be obtained using Tables 6.2 and 6.3; they are calculated from the coupling coefficients of the group D_{3d} given in Ref. [100]. Despite the time–reversal symmetry of the Hamiltonian, it is useful

Table 6.3 Multiplication table of operators in the group D_{3d}. This corresponds to the table of coupling coefficients of Ref. [100].

S	SS'	TT'	$\frac{1}{\sqrt{2}}(U_-U'_+ + U_+U'_-)$
T	ST'		$\frac{i}{\sqrt{2}}(U_-U'_+ - U_+U'_-)$
U_+	SU'_+	$-iTU'_+$	$U_-U'_-$
U_-	SU'_+	iTU'_-	$U_+U'_+$
V	SV'		
W	SW'		
X_+	SX'_+		
X_-	SX'_-		

to divide it into contributions with a well–defined transformation behaviour of the operators Ξ under time–reversal symmetry; these will be denoted by indices (s) and (as). Then the contribution of the valence band to $\mathcal{H}$ is given by:

$$\mathcal{H}^{(v)}_s = [a_1 + a_2 L_z \sigma_z]S + c(L_-U_+ + L_+U_-) \tag{6.17}$$

$$\mathcal{H}^{(v)}_{as} = +\bar{a}(L_-\sigma_+ + L_+\sigma_-)S + [\bar{b}_1\sigma_z + \bar{b}_2L_z + \bar{b}_3(L_-\sigma_+ + L_+\sigma_-)]T$$
$$+\bar{c}_1(\sigma_-U_+ + \sigma_+U_-) + \bar{c}_2(L_+\sigma_+U_+ + L_-\sigma_-U_-) \tag{6.18}$$

As the conduction band states have the full cubic O_h symmetry their transformation properties can be taken directly from Ref. [157]. Here it is enough to note that because of the compatibility relation $\Gamma_4^\pm \xrightarrow{L} L_2^\pm + L_3^\pm$ the symmetry types of the D_{3d} T, $U_\pm$ and W, $X_\pm$ transform identically. Then the conduction band contribution is

$$\mathcal{H}^{(c)}_s = \alpha S \tag{6.19}$$

$$\mathcal{H}^{(c)}_{as} = \bar{\beta}(s_x U_x + s_y U_y + s_z U_z) . \tag{6.20}$$

With these general expressions, the spin–orbit interaction in the valence band, the dispersion relations of valence and conduction band and the magnetic field dependence of the valence and conduction band energies can be specified:

1. The spin–orbit interaction is of type S and has the form

$$\mathcal{H}^{(v)}_{so} = \frac{1}{2}\Delta_{so}\sigma_z L_z . \tag{6.21}$$

Here Δ_{so} represents the splitting between the valence band states $L_{4,5}$ and L_6.

2. The form of the Zeeman energy of valence and conduction band is determined by the symmetry type T and $U_\pm$ of the magnetic field B which belongs to the class $\hat{K}_-$ of time–reversal symmetry to be

$$\mathcal{H}^{(c,ze)}(B) = \frac{1}{2}g_c\mu_B s \cdot B \tag{6.22}$$

for the conduction band and

$$\begin{aligned}\mathcal{H}^{(v,ze)}(B) &= [\overline{b}_1^{ze}\sigma_z + \overline{b}_2^{ze}L_z + \overline{b}_3^{ze}(L_-\sigma_+ + L_+\sigma_-)]B_z \\ &+ \overline{c}_1^{ze}(\sigma_-B_+ + \sigma_+B_-) + \overline{c}_2^{ze}(L_+\sigma_+B_+ + L_-\sigma_-B_-)\end{aligned} \tag{6.23}$$

for the valence band. Here g_c denote the conduction band g factor and μ_B the Bohr magneton. The parameters $\overline{b}_1^{ze}, \overline{b}_2^{ze}, \overline{b}_3^{ze}, \overline{c}_1^{ze}, \overline{c}_2^{ze}$ are constants that can be related to the effective hole g–factors. The factor $\frac{1}{2}$ comes from the definition (6.10) of the Pauli spin matrices.

3. The diamagnetic contribution to the Hamiltonian, which is proportional to B^2, is given by

$$\begin{aligned}\mathcal{H}^{(c,dia)}(B) &= \alpha^{dia}B^2 \tag{6.24}\\ \mathcal{H}^{(v,dia)}(B) &= a_{1\parallel}^{dia}B_z^2 + a_{1\perp}^{dia}(B_x^2 + B_y^2) \\ &+ (a_{2\parallel}^{dia}B_z^2 + a_{2\perp}^{dia}(B_x^2 + B_y^2))\sigma_z L_z \tag{6.25}\\ &+ c^{dia}(L_-B_-^2 + L_+B_+^2)\end{aligned}$$

for conduction and valence band.

4. The dispersion relations can be derived immediately from the symmetry type W and $X_\pm$ and the time–reversal class $\hat{K}_-$ of the wavevector. For the conduction band they are of isotropic form

$$\mathcal{H}^{(c)}(k) = E_c(0) + \frac{\hbar^2 k^2}{2m_e} \tag{6.26}$$

while for the valence band the following relation results

$$\begin{aligned}\mathcal{H}^{(v)}(k) &= E_v(k_L) + \gamma_{1\parallel}k_z^2 + \gamma_{1\perp}(k_x^2 + k_y^2) \\ &+ [\gamma_{2\parallel}k_z^2 + \gamma_{2\perp}(k_x^2 + k_y^2)]\sigma_z L_z \tag{6.27}\\ &+ \gamma_3(L_+k_+^2 + L_-k_-^2) .\end{aligned}$$

This relation shows that for the $L_{4,5}^-$ valence band states the constant energy surfaces are rotation ellipsoids around the [111]-direction. For the L_6^- states they are of rather complicated character but not of interest here. The effective masses of the uppermost valence band at the

Table 6.4 Effective masses for AgBr. In the upper line the polaron masses are given, the lower line shows the bare band masses calculated from these data. From [158].

	m_e	$m_{h\|\|}$	$m_{h\perp}$
polaron	0.288	1.79	0.79
band	0.211	1.25	0.52

L point are known for AgBr, and for the Γ_6^+ conduction band for both AgCl and AgBr from cyclotron resonance measurements. The masses for the hole in AgCl could not be measured up to now due to the extremely short lifetime of holes caused by impurity trapping. Due to the strong interaction of both electrons and holes with long–wavelength LO phonons by the Fröhlich interaction, the polaron masses differ considerably from the bare band masses, as can be seen from Table 6.4 where both the polaron masses and the bare band masses are given for AgBr.

Since the operator description of the effective Hamiltonian is rather abstract, the explicit matrix representation of the valence band Hamiltonian for the diamagentic shift, the spin–orbit interaction and the Zeeman effect are given in Tables 6.5 and 6.6.

Table 6.5 Effective valence band Hamiltonian for the diamagnetic shift.

$\phi_+^3 \alpha_v$	$\phi_-^3 \alpha_v$	$\phi_+^3 \beta_v$	$\phi_-^3 \beta_v$
$(a_{1\|\|}^{\mathrm{dia}} + a_{2\|\|}^{\mathrm{dia}})B_z^2 +$ $(a_{1\perp}^{\mathrm{dia}} + a_{2\perp}^{\mathrm{dia}})(B_x^2 + B_y^2)$	0	$c^{\mathrm{dia}}B_+^2$	0
0	$(a_{1\|\|}^{\mathrm{dia}} - a_{2\|\|}^{\mathrm{dia}})B_z^2 +$ $(a_{1\perp}^{\mathrm{dia}} - a_{2\perp}^{\mathrm{dia}})(B_x^2 + B_y^2)$	0	$c^{\mathrm{dia}}B_+^2$
$c^{\mathrm{dia}}B_-^2$	0	$(a_{1\|\|}^{\mathrm{dia}} - a_{2\|\|}^{\mathrm{dia}})B_z^2 +$ $(a_{1\perp}^{\mathrm{dia}} - a_{2\perp}^{\mathrm{dia}})(B_x^2 + B_y^2)$	0
0	$c^{\mathrm{dia}}B_-^2$	0	$(a_{1\|\|}^{\mathrm{dia}} + a_{2\|\|}^{\mathrm{dia}})B_z^2 +$ $(a_{1\perp}^{\mathrm{dia}} + a_{2\perp}^{\mathrm{dia}})(B_x^2 + B_y^2)$

Table 6.6 Effective valence band Hamiltonian for the spin–orbit and the Zeeman interaction.

$\phi_+^3 \alpha_v$	$\phi_-^3 \alpha_v$	$\phi_+^3 \beta_v$	$\phi_-^3 \beta_v$
$(\overline{b}_1^{ze} + \overline{b}_2^{ze})B_z + \frac{1}{2}\Delta_{so}$	$\overline{c}_1^{ze} B_-$	0	$\overline{c}_2^{ze} B_+$
$\overline{c}_1^{ze} B_+$	$(-\overline{b}_1^{ze} + \overline{b}_2^{ze})B_z - \frac{1}{2}\Delta_{so}$	$\overline{b}_3^{ze} B_z$	0
0	$\overline{b}_3^{ze} B_z$	$(\overline{b}_1^{ze} - \overline{b}_2^{ze})B_z - \frac{1}{2}\Delta_{so}$	$\overline{c}_1^{ze} B_-$
$\overline{c}_2^{ze} B_-$	0	$\overline{c}_1^{ze} B_+$	$-(\overline{b}_1^{ze} + \overline{b}_2^{ze})B_z + \frac{1}{2}\Delta_{so}$

6.2.3 The $\Gamma_6^+ \otimes L_{4,5}^-$ Exciton

In this section we concentrate on the $\Gamma_6^+ \otimes L_{4,5}^-$ exciton states, which form the energetically lowest electronic excitation of the silver halide crystal. As the resonant intermediate state its properties dominate the light scattering processes to be discussed later. The $\Gamma_6^+ \otimes L_6^-$ exciton states are shifted by the spin–orbit splitting Δ_{so} to higher energies. For AgBr the splitting is large compared to the exciton binding energy of about 30 meV and the mixing between the $L_{4,5}$ and L_6 hole states can be neglected in the Hamiltonian. In AgCl, however, the splitting is small and a diagonalization including the L_6 term has to be done in order to accurately describe the exciton states in this material.

As follows from the decomposition

$$\Gamma_6^+ \otimes L_{4,5}^- = 2L_3^- \tag{6.28}$$

the $\Gamma_6^+ \otimes L_{4,5}^-$ exciton shows a fourfold degenerate $1S$ ground state that can be represented by the following electron–hole pair states

$$\Phi_1^0 = -\frac{i}{\sqrt{2}}(X_L + iY_L)\alpha_v \otimes S_c\alpha_c , \tag{6.29a}$$

$$\Phi_2^0 = -\frac{i}{\sqrt{2}}(X_L + iY_L)\alpha_v \otimes S_c\beta_c , \tag{6.29b}$$

$$\Phi_3^0 = \frac{i}{\sqrt{2}}(X_L - iY_L)\beta_v \otimes S_c\alpha_c , \tag{6.29c}$$

$$\Phi_4^0 = \frac{i}{\sqrt{2}}(X_L - iY_L)\beta_v \otimes S_c\beta_c . \tag{6.29d}$$

Here the states $\Phi_{2,3}$ have a z component of the total angular momentum of $m_J = \pm 2$, while the states $\Phi_{1,4}$ have $m_J = \pm 1$. In these relations the index ^{0} denotes the states without external fields.

The degeneracy of these exciton states is lifted by the exchange interaction between electron and hole. For indirect excitons this contains only contributions due to the short–range crystal potential as no long–range dipole–dipole interaction exists in this case [94]. Therefore the indirect exciton states do not show any longitudinal-transverse splitting like in the direct case, and the value of the exchange energy is rather small. The general form of the exchange interaction for the $\Gamma_6^+ \otimes (L_{4,5}^- + L_6^-)$ state can be derived by the invariant method explained above. As the energies are expected to be small it is enough to take into account the part that is independent of external parameters. This is given by

$$\mathcal{H}^{(ex)} = \Delta_1 \sigma_z s_z + \Delta_2 (\sigma_+ s_- + \sigma_- s_+) . \qquad (6.30)$$

As usual the parameters $\Delta_{1,2}$ have non–zero values only for states with spin–singlet character [157]. To classify the exciton states according to their spin character the L–S coupling scheme has to be applied, coupling the single particle spins to the total spin $S = s - \sigma$. Here this results in three spin–triplet states with $S = 1$

$$\left.\begin{array}{c} \alpha_c \beta_v \\ \frac{1}{\sqrt{2}}(\alpha_c \alpha_v - \beta_c \beta_v) \\ \beta_c \alpha_v \end{array}\right\} \qquad\qquad S = 1 \quad (6.31)$$

and one state with $S = 0$

$$\frac{1}{\sqrt{2}}(\alpha_c \alpha_v + \beta_c \beta_v) \qquad\qquad S = 0 \quad (6.32)$$

Comparing these results to Eq. (6.29a) shows that the states $\Phi_{2,3}$ are pure spin states with $S = 1$ and therefore have an exchange energy of zero. The states $\Phi_{1,4}$ are linear combitions of spin singlet and triplet states. According to Eq. (6.30) the states $\Phi_{1,4}$ remain degenerate even with inclusion of the exchange energy and are shifted by $2\Delta_1 = \Delta$ to higher energies compared to the degenerate triplet states $\Phi_{2,3}$.

To describe the behaviour of the $\Gamma_6^+ \otimes L_{4,5}^-$ exciton states in a magnetic field we consider both the exchange interaction and the linear Zeeman terms. The additional inclusion of a diamagnetic shift is straightforward (see [159]). In order to have the most elegant description we introduce effective hole spin–operators $\sigma_z^{eff}, \sigma_+^{eff}, \sigma_-^{eff}$ by the following definition

$$\sigma_z^{eff} = \mathcal{P}_{4,5}\sigma_z \qquad\qquad (6.33a)$$
$$\sigma_\pm^{eff} = \mathcal{P}_{4,5}L_\pm\sigma_\pm . \qquad\qquad (6.33b)$$

Here $\mathcal{P}_{4,5}$ denotes a projection operator onto the $L_{4,5}$ states. The zero of the energy scale is chosen to be at the energy of the singlet–triplet states. Introducing effective g factors by

$$g_v^{\parallel} = 2(\overline{b_1^{ze}} + \overline{b_2^{ze}})/\mu_B \qquad\qquad (6.34)$$
$$g_v^{\perp} = i\sqrt{2}\overline{c_2^{ze}}/\mu_B \qquad\qquad (6.35)$$

Table 6.7 Effective exciton Hamiltonian: Exchange interaction and linear Zeeman energy

ϕ_1^0	ϕ_2^0	ϕ_3^0	ϕ_4^0
$\frac{1}{2}(g_c - g_v^{\parallel})B_z$	$\frac{1}{2}g_c(B_x - iB_y)$	$-\frac{1}{2}g_v^{\perp}(B_y + iB_x)$	0
$\frac{1}{2}g_c(B_x + iB_y)$	$-\frac{1}{2}(g_c + g_v^{\parallel})B_z - \Delta$	0	$-\frac{1}{2}g_v^{\perp}(B_y + iB_x)$
$-\frac{1}{2}g_v^{\perp}(B_y - iB_x)$	0	$\frac{1}{2}(g_c + g_v^{\parallel})B_z - \Delta$	$\frac{1}{2}g_c(B_x - iB_y)$
0	$-\frac{1}{2}g_v^{\perp}(B_y - iB_x)$	$\frac{1}{2}g_c(B_x + iB_y)$	$-\frac{1}{2}(g_c - g_v^{\parallel})B_z$

the effective Hamiltonian for the exciton states (Eq. (6.29a)) can be written as

$$\mathcal{H} = \frac{1}{2}\Delta(\sigma_z^{\text{eff}}s_z - 1) + \frac{1}{2}g_c\mu_B s \cdot B$$
$$- \frac{1}{2}g_v^{\parallel}\mu_B\sigma_z^{\text{eff}}B_z - \frac{1}{2}g_v^{\perp}(\sigma_x^{\text{eff}}B_y - \sigma_y^{\text{eff}}B_x) \, . \tag{6.36}$$

Comparing this result to earlier works [160, 159] an additional term occurs describing the hole Zeeman effect in transverse magnetic fields with a g–factor $g_v^{\perp}$. Obviously the analogy of the L point symmetry to the wurtzite structure that has led to the assumption that this term may be neglected is not correct as is shown here by a complete symmetry analysis.

The contributions proportional to $g^{\perp}$ in the effective Hamiltonian have direct consequences as they mix all four exciton states under the action of a magnetic field. This mixing strongly influences the strength of the optical absorption of the exciton states as will be discussed in detail in the next section. The Zeeman energy splitting of the exciton is also different in the exact model as can be seen from the result of a numerical diagonalization. In Fig. 6.2 the dependence of the exciton energies on a magnetic field along the [001] crystallographic direction is shown for $g_v^{\perp} = 0$ and all other parameters taken from Ref. [159]. The characterisitic feature is the crossing between the states 1 and 3, which do not mix in this model. In Fig. 6.3 the results are shown assuming $g_v^{\perp} = 1$ and the other parameters taken from the analysis of the quantum beat experiments discussed later in this chapter. While the states 4 and 2 show a similar dependence to the case of $g_v^{\perp} = 0$, no crossing of states 1 and 3 occurs due to their mixing induced by the magnetic field. A straightforward calculation of the magnetic field dependence of the intensity of the 2TO(L) scattering process, the details of which will not be discusssed here, but which was the essential point of the analysis in Ref. [159], shows an even better agreement with the experimental data of this work.

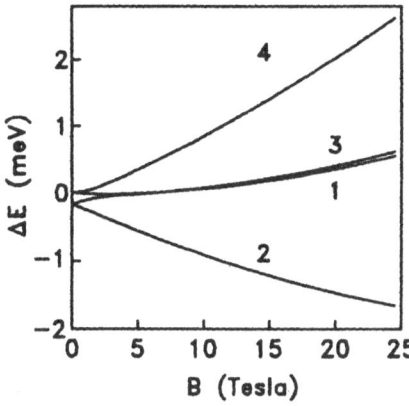

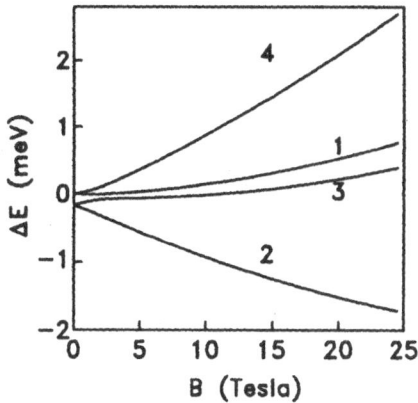

Fig. 6.2 Energy splitting of the $\Gamma_6^+ \otimes L_{4,5}^-$ exciton as a function of a magnetic field along the [001] direction. Parameters: $g_c = 1.46, g_v^{\parallel} = 2.61, g_v^{\perp} = 0, \Delta = 0.17$ meV. The numbers denote the exciton states

Fig. 6.3 Same as Fig. 6.2, but with the full Zeeman energy. Parameters: $g_c = 1.45, g_v^{\parallel} = 2.6, g_v^{\perp} = 1, \Delta = 0.15$ meV.

6.3 Matrix Elements of the Indirect Transition

In this section, the dependence of the strength of the indirect transition to the $\Gamma_6^+ \otimes L_{4,5}^-$ exciton states on the polarization of the light field will be derived from a group–theoretical analysis. According to Eq. (4.54) the polarization selection rules are governed by an effective transition dipole moment $M_{j,l}^{\eta}$. From this the coupling matrices of the exciton states with the light field can be derived for the basic Raman scattering process resonant with this exciton, which according to Fig. 6.4 consists of an absorption followed by an emission process both involving a momentum conserving L point phonon. All experimental results [155, 42] have shown that the TO(L) phonon has the

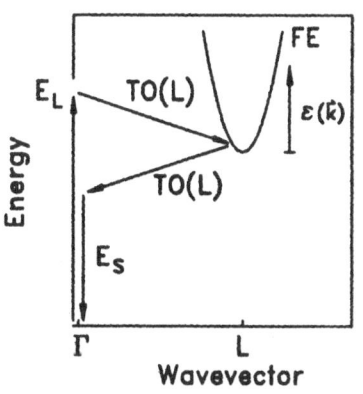

Fig. 6.4
Schematic representation of the 2TO(L) scattering process.
E_L: laser photon energy; E_S: scattered photon energy; $\epsilon(\mathbf{k})$: kinetic exciton energy.

Table 6.8 Symmetry of L and X phonons in crystals with NaCl structure. (After [94]). In AgBr the eigenvectors of the TA(L) and TO(L) phonons are exchanged [161].

phonon	symmetry	phonon	symmetry
LA(L)	L_1^+	LA(X)	X_4^-
TA(L)	L_3^+	TA(X)	X_5^-
LO(L)	L_1^-	LO(X)	X_4^-
TO(L)	L_3^-	TO(X)	X_5^-

strongest transition strength. However, it has to be noted that, according to neutron scattering experiments, in AgBr an exchange of the eigenvectors of the TO(L) and TA(L) modes occurs [161]. To keep the following discussion as general as possible we will use the notation TA(L) mode. In the discussion of the experimental results in AgBr the notation TO(L) will be used again.

6.3.1 Transition Moments: No Spin–Orbit Interaction

For the group–theoretical analysis the case of vanishing spin–orbit interaction will be considered first. This is possible as the electron–phonon interaction is to first order spin independent. For this purpose we need the symmetries of the L point phonons that are given in Table 6.8, together with those of the X phonons needed later.

The number of linearly independent matrix elements follows directly from the multiplicity of the unit representation in the direct product of the representations [94]

$$\mathcal{D}^*_{\text{valence band}} \otimes \mathcal{D}_{\text{conduction band}} \otimes \mathcal{D}_{\text{dipole operator}} \otimes \mathcal{D}_{\text{phonon}}$$

From the relations

$$L_3^- \otimes \Gamma_1^+ \otimes \Gamma_4^- \otimes L_1^\pm = L_1^\pm + L_2^\pm + 2L_3^\pm \tag{6.37}$$

$$L_3^- \otimes \Gamma_1^+ \otimes \Gamma_4^- \otimes L_3^\pm = 2L_1^\pm + 2L_2^\pm + 4L_3^\pm \tag{6.38}$$

it follows immediately that the scattering is forbidden with LO(L) and TO(L) phonons, while it is allowed for the TA(L) and LA(L) modes. Thereby the LA(L) scattering can be described by *one*, and the TA(L) scattering by *two* linearly independent matrix elements.

In order to proceed further, intermediate states of scattering via the valence and conduction bands which are energetically near the indirect transition because of the energy denominators in Eq. (4.54) have to be considered. By symmetry the following states are possible

Table 6.9 Reduced transition matrix elements for the indirect exciton in AgBr for valence band scattering. x_l, y_l, z_l are the polarization vectors referred to the coordinates of Table 6.1, M_0 is the reduced dipole matrix element and $\epsilon^{TA}, \eta^{TA}, \epsilon^{LA}$ are the reduced matrix elements of the valence band scattering with TA(L) and LA(L) phonons. For TA(L) scattering the two lines denote the coupling with the two degenerate phonon modes $(+,-)$.

	x_l	y_l	z_l
Φ_+^3	$-M_0\epsilon^{LA}$	$-iM_0\epsilon^{LA}$	0
Φ_-^3	$M_0\epsilon^{LA}$	$-iM_0\epsilon^{LA}$	0
Φ_+^3	$M_0\epsilon^{TA}$ 0	$-iM_0\epsilon^{TA}$ 0	0 $-M_0\eta^{TA}$
Φ_-^3	0 $-M_0\epsilon^{TA}$	0 $-iM_0\epsilon^{TA}$	$M_0\eta^{TA}$ 0

LA(L) scattering :

conduction band : $\Gamma_1^+ \rightarrow L_1^+$ valence band : $\Gamma_4^- \rightarrow L_3^-$,

TA(L) scattering (in AgBr : TO(L)–mode):

conduction band : $\Gamma_1^+ \rightarrow L_3^+$ valence band : $\Gamma_4^- \rightarrow L_3^-$.

Using the table of coupling coefficients for the symmetry groups O_h and D_{3d} [100] the reduced matrix elements given in Table 6.9 for valence band scattering and in Table 6.10 for conduction band scattering are obtained. Here one has to take into account that, because of the relation $\Gamma_4^- \rightarrow L_2^- + L_3^-$, at the Γ point the dipole transition $(\Gamma_4^- \rightarrow \Gamma_1^+)$ is determined by *one* reduced matrix element, but at the L point $(L_3^- \rightarrow L_1^+ + L_3^+)$, however, by *two* matrix elements.

6.3.2 Transition Moments: Spin–Orbit Interaction

To obtain the total transition moments for the indirect exciton states, the contributions of the various intermediate states have to be summed up. Here it is essential to take the spin–orbit interaction into account. As most important intermediate states (see the band structure in Fig. 6.1) the Γ_8^-, Γ_6^- in the valence band and $L_{4,5}^+ + L_6^+$ in the conduction band are included. Then the following transition moments result

Table 6.10 Reduced transition matrix elements for the indirect exciton in AgBr for conduction band scattering. All abbreviations as in Table 6.9, except M_1, M_2, M_3 which denote the dipole transition matrix elements at the L point of the BZ.

	x_l	y_l	z_l
Φ^3_+	$-\dfrac{1}{2}M_1\delta^{LA}$	$\dfrac{i}{2}M_1\delta^{LA}$	0
Φ^3_-	$\dfrac{1}{2}M_1\delta^{LA}$	$\dfrac{i}{2}M_1\delta^{LA}$	0
Φ^3_+	$-\dfrac{1}{2}M_3\delta^{TA}$ 0	$\dfrac{i}{2}M_3\delta^{TA}$ 0	0 $\dfrac{i}{\sqrt{2}}M_2\delta^{TA}$
Φ^3_-	0 $\dfrac{1}{2}M_3\delta^{TA}$	0 $\dfrac{i}{2}M_3\delta^{TA}$	$-\dfrac{i}{\sqrt{2}}M_3\delta^{TA}$ 0

$$
M^\eta_{i,l} = \sum_{\Gamma_8^-} \frac{<\psi^6_i(\Gamma)|e\cdot p|\phi^8_\lambda(\Gamma)><\phi^8_\lambda(\Gamma)|V^\eta|\phi^3_i(L)>}{E_v(L_3^-) - E_\lambda(\Gamma_8^-) \mp \hbar\omega^\eta} \tag{6.39}
$$
$$
+ \sum_{\Gamma_6^-} \frac{<\psi^6_i(\Gamma)|e\cdot p|\phi^6_\lambda(\Gamma)><\phi^6_\lambda(\Gamma)|V^\eta|\phi^3_i(L)>}{E_v(L_3^-) - E_\lambda(\Gamma_6^-) \mp \hbar\omega^\eta}
$$
$$
+ \sum_{L_{4,5,6}} \frac{<\psi^6_i(\Gamma)|e\cdot p|\psi^{4,5,6}_\lambda(L)><\psi^{4,5,6}_\lambda(L)|V^\eta|\phi^3_i(L)>}{E_c(\Gamma_6^+) - E_c(L_{4,5,6}^+) \mp \hbar\omega^\eta} \; .
$$

Here $\psi^6_i(\Gamma)$ and $\phi^3_i(L)$ denote the electron and hole contributions to the indirect exciton states Φ_i, while $\phi^{6,8}_\lambda(\Gamma)$ are the Bloch functions of the valence band intermediate states and $\psi^{4,5,6}_\lambda(L)$ those of the conduction band. For the TA(L) phonon scattering the matrix element has to be calculated for each phonon mode (here denoted by + and −) separately, as the scattering processes have to be summed incoherently over the phonon modes.

The results of the calculation using the states from Sect. 6.2 is given in Table 6.11 for TA(L) and LA(L) scattering. Here the following abbreviations have been introduced:

$$
\sigma^{TA}_S = \frac{\sqrt{2}\xi S^{TA} - i(2+\beta)P^{TA}}{(1+2\beta)Q^{TA} + \xi R^{TA}} \tag{6.40}
$$
$$
\sigma^{TA}_T = \frac{(1-\beta)P^{TA}}{(1+2\beta)Q^{TA} + \xi R^{TA}} \tag{6.41}
$$

Table 6.11 Total transition moments of the indirect exciton states for TA(L) and LA(L) scattering. Further explanation see text.

exciton state	TA(+)	TA(−)	LA
$M^{\eta}_{1,l}$	$(x_l - iy_l)$	$-i\sigma^{TA}_S$	$-(x_l + iy_l)$
$M^{\eta}_{2,l}$	$\sigma'^{TA}_T z_l$	$\sigma^{TA}_T(x_l + iy_l)$	0
$M^{\eta}_{3,l}$	$\sigma^{TA}_T(x_l - iy_l)$	$\sigma'^{TA}_T z_l$	0
$M^{\eta}_{4,l}$	$i\sigma^{TA}_S z_l$	$-(x_l + iy_l)$	$(x_l - iy_l)$

$$\sigma'^{TA}_T = \frac{2(1 - \beta)Q^{TA}}{(1 + 2\beta)Q^{TA} + \xi R^{TA}} \tag{6.42}$$

with

$$Q^{TA} = \frac{1}{3}M_0\epsilon^{TA} \quad P^{TA} = \frac{1}{3}M_0\eta^{TA} \quad R^{TA} = \frac{1}{2}M_3\delta^{TA} \quad S^{TA} = \frac{1}{2}M_2\delta^{TA}. \tag{6.43}$$

The quantities β and ξ denote the following ratios of energy differences

$$\beta = \frac{E_v(L^-_{4,5}) - E_v(\Gamma^-_8)}{E_v(L^-_{4,5}) - E_v(\Gamma^-_6)} \tag{6.44}$$

$$\xi = \frac{E_v(L^-_{4,5}) - E_v(\Gamma^-_8)}{E_c(L^+_{4,5,6}) - E_c(\Gamma^+_6)}. \tag{6.45}$$

The derived relations show that for LA(L) scattering the transition moments are of a relatively simple structure due to the L^+_1 phonon symmetry and are pure σ transitions with moments only perpendicular to the threefold rotation axis of the D_{3d}. For the TA(L) scattering the moments are much more complex with contributions both parallel (π) and perpendicular (σ) to the D_{3d} axis. In particular, depending on the ratio of valence to conduction band scattering the parameter σ^{TA}_S can also acquire a complex value. Due to the spin–orbit interaction, the spin triplet states can also be excited optically via the TA(L) process. In order to estimate this effect and to obtain order of magnitudes for the parameters $\sigma^{TA}_S, \sigma^{TA}_T, \sigma'^{TA}_T$ for AgBr, we assume that the following relation, valid for full O_h symmetry, holds

$$\frac{M_0}{\sqrt{3}} = M_2 = M_3, \qquad \epsilon^{TA} = \eta^{TA} = \delta^{TA}. \tag{6.46}$$

From the energies of the direct and indirect exciton states

$$E_{gx}(\Gamma^-_8) = 4.25 \text{ eV}, \ E_{gx}(\Gamma^-_6) = 4.85 \text{ eV}, \ E_{gx}(L^-_{4,5}) = 2.684 \text{ eV}$$

the exciton binding energies

$$E_{bx}(\Gamma) = 0.3 \text{ eV}, \ E_{bx}(L) = 0.028 \text{ eV}$$

and the energies of the conduction band states

$$E(\Gamma_6^+) - E(\Gamma_{4,5,6}^+) \simeq 5 \text{ eV},$$

which can be approximately taken from the band structure (Fig. 6.1), it follows that

$$\beta = 0.75 \text{ and } \xi = 0.37 \ .$$

Then we have for the TA(L) process

$$\sigma_S^{TA} = 1.0 + 0.16i \ , \ \sigma_T^{TA} = 0.08 \text{ and } \sigma_T'^{TA} = 0.16 \ .$$

The transition moments are therefore complex with π and σ components of equal magnitude. The absorption into the pure triplet states, however, is about two orders of magnitude smaller than the singlet absorption. This explains why this contribution has not been found experimentally up to now. More accurate, and somewhat different, values for these parameters can be derived from the experiments discussed in the next section.

6.4 Time–Resolved Light Scattering in AgBr: The 2TO(L) Process

The topic of this section is the dominant process of light scattering resonant with the indirect exciton in AgBr involving two TO(L) phonons. Higher order scattering will be postponed until the next section. After discussing the experimental results for the time dependence of the scattered intensity and the investigations of quantum beats in this process, an analysis using the methods developed in Chap. 5 is presented. For a first, more qualitative discussion, the phenomenological density matrix treatment of Sect. 5.2 will be used. Quantitative results then can be obtained from the transform–limited theory of Sect. 5.3.

The scattered intensity of the two–phonon process is determined by the transiton moments $M_{i,l}^\eta$. Therefore it depends on

- the orientation of the crystallographic axes
- the polarization state of the exciting and of the scattered light.

In the experiments a 90° scattering geometry as shown in Fig. 6.5 was employed. As a fixed coordinate system where the experiments are described we have chosen the following directions

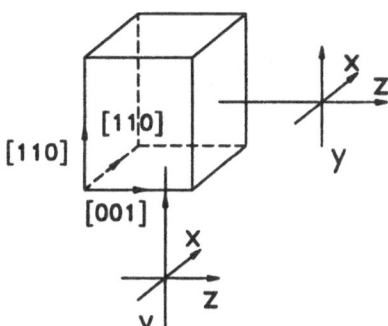

Fig. 6.5
90°–scattering geometry for the measurements with the indirect exciton in AgBr. x, y, z: fixed coordinate system related to the cubic crystal axis.

$$x = \frac{1}{\sqrt{2}}(1,-1,0) \qquad y = \frac{1}{\sqrt{2}}(1,1,0) \qquad z = (0,0,1) . \qquad (6.47)$$

The exciting laser beam was directed along the y axis ($\mathbf{k}_L \| \mathbf{y}$), the scattered light was observed in the z direction ($\mathbf{k}_S \| \mathbf{z}$). To specify the polarization state the Poincaré number [162] χ was found to be very convienient. Therefore we have

$$e_L = \frac{1}{\sqrt{1 + |\chi_L|^2}} (\mathbf{x} + \chi_L \mathbf{z}), \qquad (6.48)$$

for the excitation and

$$e_S = \frac{1}{\sqrt{1 + |\chi_S|^2}} (\mathbf{x} + \chi_S \mathbf{y}) \qquad (6.49)$$

for the detection of the scattered light.

6.4.1 Experimental Results

According to the results of previous studies [22] the measurement of the time–dependence of the 2TO(L) scattering process allows one to determine the excitonic lifetime and thus the excitonic relaxtion processes directly. However, in this section we will concentrate on the possibility offered by polarized, transform–limited light scattering in connection with the quantum beat method to explore coherent and incoherent dynamics of the indirect exciton states. Most important for these measurements is the spectral separation of the relevant 2TO(L) scattering line from the other light scattering processes. This can be done by applying the maximum spectral resolution that is compatible with the required time–resolution, i.e. to work under transform–limited conditions. The investigations will be restricted to the range of small exciton kinetic energies as here one can expect long relaxation times and, hence, a strong impact of coherent processes on the light scattering.

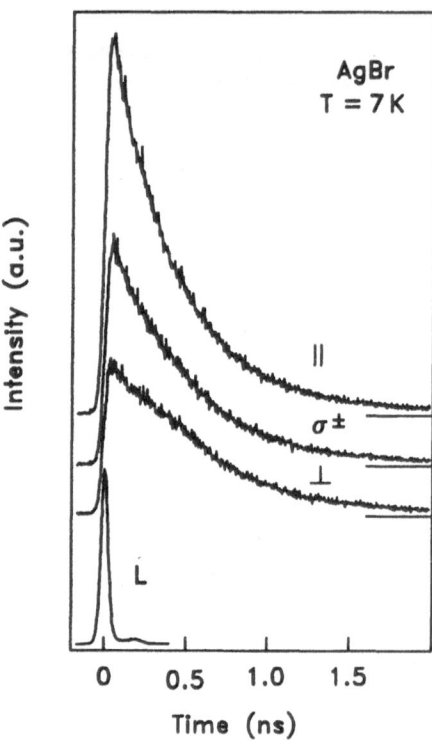

Fig. 6.6
Temporal behaviour of the intensity of the 2TO(L) scattering process for different detection polarization directions. L marks the system
response to a 5 ps laser pulse, the
lines at the right–hand side mark
the zero of intensity of each curve.
Excitation polarization: $e_L \| x$. Po-
larization directions of the scattered
light: $\|$: $e_S \| x$, $\perp$: $e_S \| y$, $\sigma^{\pm}$: left and
right circular polarization.

6.4.1.1 Polarization Dependence.
As shown by the results of Chap. 5,
in principle it should be possible by suitable choices of the polarization states
of the excitation laser pulses and of the scattered light to obtain from the
analysis of the time dependence both the characteristic relaxation rates of
the exciton dynamics and the transition matrix elements.

An example of this type of measurement is shown in Fig. 6.6. Here the
excitation photon energy was near the band minumum at an exciton kinetic
energy of $\epsilon = 0.8$ meV. The excitation was linearly polarized along the x
axis, the detection was both linearly and circularly polarized. The differences
between the various polarization directions are clearly visible. While the com-
ponent with polarization parallel to that of excitation ($\|$) shows a rather fast
decay, the component with perpendicular polarization direction ($\perp$) decays
in a non–exponential way. The left and right circularly polarized intensities
exhibit a similar time dependence being otherwise identical in their decay
behaviour. This can be seen even more clearly in the decay curves shown in
Fig. 6.7 in a magnetic field of $B = 0$ T and excited at a somewhat lower
photon energy. For all measurements rather long excitonic decay times of the
order of 200 to 400 ps can be estimated.

Most interesting is the fast rise of the scattered intensity of the parallel
component for early times, which is almost identical to that of the laser pulse.
This already implies the influence of the excitation and detection on the time–

resolved scattered intensities as, according to the discussion in Sect. 5.3, this is the fingerprint of an indirect exciton state in the temporal behaviour of a Raman scattering process in the transform–limited theory.

Neither the non–exponential decay nor the fast rise of the scattered intensity can be explained in the usual treatment of time–dependent processes by rate equations (see e.g. [22]).

Obviously, the fast Raman contribution is also present in the perpendicular component of the scattered intensity. In the previous work on time–resolved light scattering in AgBr(see e.g. [23]) it remained an open question whether this is due to the selection rules of the indirect transition or is induced by a depolarization of the exciting laser pulse in the sample. Here we will show that it is indeed the peculiar form of the indirect transition moments that leads to this effect.

6.4.1.2 Quantum Beat Spectroscopy. Further information on the dynamics of the indirect excitons is available from quantum–beating in the scattering process, as this is a clear indication of the coherence of the exciton states. A typical set of measurements showing the quantum beats is displayed in Fig. 6.7. At zero field strength $B = 0$ T they exhibit an identical

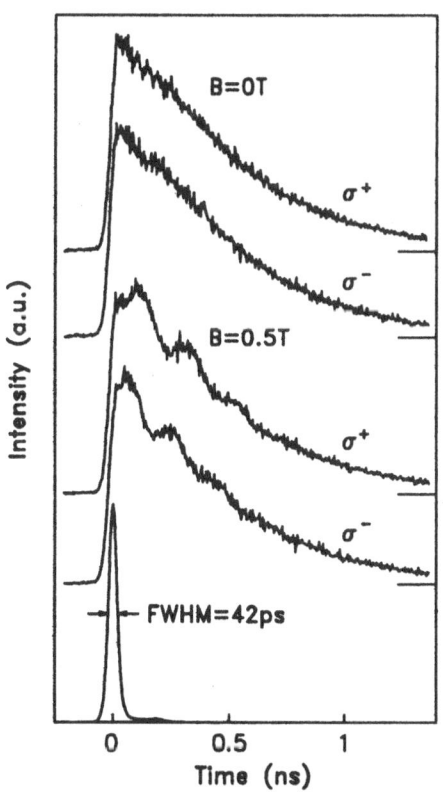

Fig. 6.7

Time dependence of the intensity of the 2TO(L) scattering process without magnetic field (upper curves) and applying a magnetic field of $B = 0.5$ T after excitation by laser pulses of 8 ps duration with a photon energy of $E_L = 2.6931$ eV and plane polarized parallel to the [$1\bar{1}0$] direction. The scattered light was detected with left (σ^+) and right circular (σ^-) polarization. $T = 2$ K. The lower curve shows the temporal system response to the laser pulses.

time dependence for both left and right circular polarization, already known from the last section. On applying a magnetic field along the [001] crystallographic direction pronounced oscillations in the temporal decay occur, the quantum beats, the phase of which is different by approximately 180° in the two components. Most surprising, however, is that the modulation degree of the signals, i.e. the ratio of the quantum beat amplidute and the back–ground, is rather small despite the long dephasing times of the order of 200 – 300 ps. As this small ratio was found also for other detection polarization directions, e.g. for linear polarization parallel and perpendicular to the excitation polarization [163], this effect has to be correlated to the matrix elements of the indirect transition. In the next section we will thus apply the density matrix treatment to analyse the experimental results and to gain at least qualitative insight.

6.4.2 Density Matrix Treatment

In the first step one has to determine the irreducible components of the density matrix of the excitonic system. Thereby each inequivalent L point has to be taken into acount and the final result is obtained by an incoherent summation over all L points and the two phonon modes. The latter corresponds to the occurrence of four different scattering processes that may be denoted by the symbols $++$, $--$, $+-$ and $-+$.

In order to keep the discussion as simple as possible, only the strongly absorbing singlet–triplet (S–T) mixed states $\Phi_{1,4}^0$ will be taken into account and the pure triplet states (T) are completely neglected. This restricts the treatment to small magnetic fields where the mixing of the S–T and pure T states can be neglected. In this range of fields, which actually goes up to about 0.5 T, we may in addition assume that the excitonic relaxation processes do not change with field.

As the S–T states transform like the representation L_3^-, the relation

$$L_3^{-*} \otimes L_3^- = L_1^+ + L_2^+ + L_3^+ \tag{6.50}$$

holds. The density matrix can therefore be decomposed into the irreducible components

$$\rho^{(1)} = \frac{1}{\sqrt{2}}(\rho_{11} + \rho_{44}) \tag{6.51a}$$

$$\rho^{(2)} = \frac{1}{\sqrt{2}}(\rho_{44} - \rho_{11}) \tag{6.51b}$$

$$\rho^{(3)} = \begin{pmatrix} \rho_{41} \\ \rho_{14} \end{pmatrix}. \tag{6.51c}$$

Accordingly, the dynamics of the exciton is fully characterized by three relaxation rates, the population relaxation rate γ_1, the cross relaxation rate γ_2 and the phase relaxation rate of the Hertzian coherence γ_3.

The energetic splitting of the states at small magnetic fields is determined by the Hamiltonian H_{ex} which is obtained by considering only the S–T states to be

$$H_{ex} = \begin{pmatrix} \frac{1}{2}(g_c - g_v^{\parallel})B_z & 0 \\ 0 & -\frac{1}{2}(g_c + g_v^{\parallel})B_z \end{pmatrix}. \tag{6.52}$$

The temporal dynamics is governed by the master equation (5.84)

$$\frac{d\rho}{dt} = -\frac{i}{\hbar}[H_{ex}, \rho] + R(t) + G(t)$$

whereby R denotes the relaxation matrix and G a $\delta(t)$–like excitation pulse. The solution of this equation is straightforward and given by

$$\rho^{(1)}(t) = G^{(1)}e^{-\gamma_1 t} \tag{6.53a}$$
$$\rho^{(2)}(t) = G^{(2)}e^{-\gamma_2 t} \tag{6.53b}$$
$$\rho^{(3)}(t) = G^{(3)}\begin{pmatrix} e^{-i\delta/t - \gamma_3 t} \\ e^{i\delta/t - \gamma_3 t} \end{pmatrix}. \tag{6.53c}$$

Here $\delta = (g_v^{\parallel} - g_c)B_z/\hbar$ denotes the splitting frequency between the states 1 and 4, and $G^{(i)}$ with $i = 1, \ldots, 3$ the decomposition of the generation matrix into the irreducible components. For $\delta \neq 0$ the irreducible component $\rho^{(3)} \neq 0$ represents a Hertzian coherence that oscillates with the angular frequency δ. By an appropriate choice of excitation and detection polarization this coherence manifests itself in the form of a quantum beat in the scattered light being damped by the relaxation rate γ_3.

Choosing the basis polarization states of excitation as $|1\rangle = |z\rangle$, $|2\rangle = |x\rangle$ and of detection as $|1\rangle = |x\rangle$, $|2\rangle = |y\rangle$ the polarization density matrices for the exciting laser pulse and for the scattered light are given by

$$\underline{\underline{I}}_L = \frac{1}{\sqrt{1 + |\chi_L|^2}}\begin{pmatrix} |\chi_L|^2 & \chi_L \\ \chi_L^* & 1 \end{pmatrix} \tag{6.54}$$

$$\underline{\underline{I}}_S = \frac{1}{\sqrt{1 + |\chi_S|^2}}\begin{pmatrix} 1 & \chi_S^* \\ \chi_S & |\chi_S|^2 \end{pmatrix}. \tag{6.55}$$

The coupling matrices of excitation and detection referred to these basis states obtained from the transition moments are given in Table 6.12. Calculating with these coupling matrices the excitation matrices and the coherency matrices of the scattered light, one finds that for real values of the ratios σ_S of the σ to π polarized transition moments there exist no coherences at all in the scattered light, neither in the form of a polarazation nor in the form of a quantum beat, as the contributions of the two phonon modes exactly cancel each other in this case. Only for a complex value of σ_S can the observed polarization effects and quantum beats be explained. A complex σ_S implies that

Table 6.12 Coupling matrices of the singlet–triplet states of the indirect exciton in silver halides for the TA(L) phonon process. The first row of each matrix gives the coupling of the state Φ_1^0, the second row that of the state Φ_4^0. Abbreviations: $a = \sigma_S\sqrt{2/3}$, $b = \sigma_S/\sqrt{3}$.

l	$\underline{M}_L(+)$	$\underline{M}_L(-)$	$\underline{M}_S(+)$	$\underline{M}_S(-)$
1	$\begin{pmatrix} \sqrt{2/3} & ib \\ -i & 0 \end{pmatrix}$	$\begin{pmatrix} -ib & -\sqrt{2/3} \\ 0 & -i \end{pmatrix}$	$\begin{pmatrix} -i & 0 \\ -1/\sqrt{3} & ia \end{pmatrix}$	$\begin{pmatrix} 0 & -i \\ -ia & 1/\sqrt{3} \end{pmatrix}$
2	$\begin{pmatrix} \sqrt{2/3} & ib \\ 1/\sqrt{3} & -ia \end{pmatrix}$	$\begin{pmatrix} -ib & -\sqrt{2/3} \\ ia & -1/\sqrt{3} \end{pmatrix}$	$\begin{pmatrix} 1/\sqrt{3} & -ia \\ -i & 0 \end{pmatrix}$	$\begin{pmatrix} ia & -1/\sqrt{3} \\ 0 & -i \end{pmatrix}$
3	$\begin{pmatrix} \sqrt{2/3} & ib \\ i & 0 \end{pmatrix}$	$\begin{pmatrix} -ib & -\sqrt{2/3} \\ 0 & i \end{pmatrix}$	$\begin{pmatrix} i & 0 \\ \sqrt{1/3} & -ia \end{pmatrix}$	$\begin{pmatrix} 0 & i \\ ia & -1/\sqrt{3} \end{pmatrix}$
4	$\begin{pmatrix} \sqrt{2/3} & ib \\ -1/\sqrt{3} & ia \end{pmatrix}$	$\begin{pmatrix} -ib & -\sqrt{2/3} \\ -ia & 1/\sqrt{3} \end{pmatrix}$	$\begin{pmatrix} -1/\sqrt{3} & ia \\ i & 0 \end{pmatrix}$	$\begin{pmatrix} -ia & 1/\sqrt{3} \\ 0 & i \end{pmatrix}$

both valence and conduction band scattering are of importance in the indirect transition. The parameter σ_S can be obtained from an accurate analysis of the degree of polarization of the 2TO(L) process. This analysis, however, depends very sensitively on the temporal behaviour of the scattered light and can therefore be done reliably in the framework of the transform–limited theory, as is undertaken in the next section.

6.4.3 Transform–Limited Theory

As the measurements at higher field will also be included in the full analysis using the transform–limited theory, the interaction of the triplet states with the light field has to be taken into account. The relevant coupling matrices for these states are given in Table 6.13.

The intensity of the Raman part of the scattering process is now given by

$$I^{RS}(t, e_L, e_S) \propto \sum_K \sum_{l=1}^{4} \sum_{\eta=+,-} \sum_{\eta'=+,-} \left| \sum_m M_{m,l}^\eta \cdot e_L (M_{m,l}^{\eta'} \cdot e_S)^* \Psi_{m,l,K}(t) \right|^2 \tag{6.56}$$

Table 6.13 Coupling matrices of the pure triplet states of the indirect exciton in AgBr. The first row gives the coupling of the state Φ_2^0, the second one that of the state Φ_3^0. Abbreviations: $a = \sigma_T \sqrt{2/3}$, $b = \sigma_T/\sqrt{3}$, $a' = \sigma_T' \sqrt{2/3}$, $b' = \sigma_T'/\sqrt{3}$.

l	$\underline{M}_L(+)$	$\underline{M}_L(-)$	$\underline{M}_S(+)$	$\underline{M}_S(-)$
1	$\begin{pmatrix} b' & a \\ 0 & i\sigma_T \end{pmatrix}$	$\begin{pmatrix} a & b' \\ -i & \sigma_T \end{pmatrix}_0$	$\begin{pmatrix} 0 & i\sigma_T \\ a' & -b \end{pmatrix}$	$\begin{pmatrix} -i\sigma_T & 0 \\ -b & a' \end{pmatrix}$
2	$\begin{pmatrix} b' & a \\ -a' & b \end{pmatrix}$	$\begin{pmatrix} a & b' \\ b & -a' \end{pmatrix}$	$\begin{pmatrix} -a' & b \\ 0 & i\sigma_T \end{pmatrix}$	$\begin{pmatrix} b & -a' \\ i\sigma_T & 0 \end{pmatrix}$
3	$\begin{pmatrix} b' & -i\sigma_T \\ 0 & a \end{pmatrix}$	$\begin{pmatrix} a & b' \\ i\sigma_T & 0 \end{pmatrix}$	$\begin{pmatrix} 0 & -i\sigma_T \\ -a' & b \end{pmatrix}$	$\begin{pmatrix} a & 0 \\ -i\sigma_T & a' \end{pmatrix}$
4	$\begin{pmatrix} b & a \\ a' & -b \end{pmatrix}$	$\begin{pmatrix} a & b' \\ -b & a' \end{pmatrix}$	$\begin{pmatrix} a' & -b \\ 0 & -i\sigma_T \end{pmatrix}$	$\begin{pmatrix} -b & a' \\ -i\sigma_T & 0 \end{pmatrix}$

where the summation is to be performed over the exciton states ($m = 1, \ldots, 4$), the L points ($l = 1, \ldots, 4$), the phonon modes ($+, -$) and the exciton wavevectors K.

The hot luminescence component is obtained as

$$I^{HL}(t, e_L, e_S) = \tag{6.57}$$

$$\frac{1}{4} \sum_K \sum_{l=1}^{4} \sum_{\eta=+,-} \sum_{K'} \sum_{l'=1}^{4} \sum_{\eta'=+,-} \sum_m \sum_{m'} \left| M_{m,l}^{\eta} \cdot e_L M_{m',l'}^{\eta'} \cdot e_S \right|$$

$$\times \gamma_L \cdot \gamma_S \cdot A_{lk,K,m,l} \cdot A_{sk,K',m',l'} \cdot A_{K,m,l,K',m',l'}(t) \ .$$

The abbreviations are explained in Eqs. (5.147)–(5.149). The functions $\Psi_{m,l,K}(t)$, $A_{lk,K,m,l}$, $A_{sk,K',m',l'}$ and $A_{K,m,l,K',m',l'}(t)$ depend on the energies of the exciton states, which change with applied magnetic field. They have been obtained by a numerical diagonalization of the Hamiltonian (Eq. (6.36)).

At a magnetic field $B = 0$ the scattered intenities depend only on the transition moments of the exciton states, i.e. the quantities σ_S, σ_T and σ_T'. At non–zero field the energies depend in addition on the g–factors and on the exchange energy Δ. In particular, these determine the quantum beat oscillation frequencies. The temporal decay of the scattered light is determined by both the coherent relaxation rate Γ and the energy relaxation rate Γ_{ER}. In

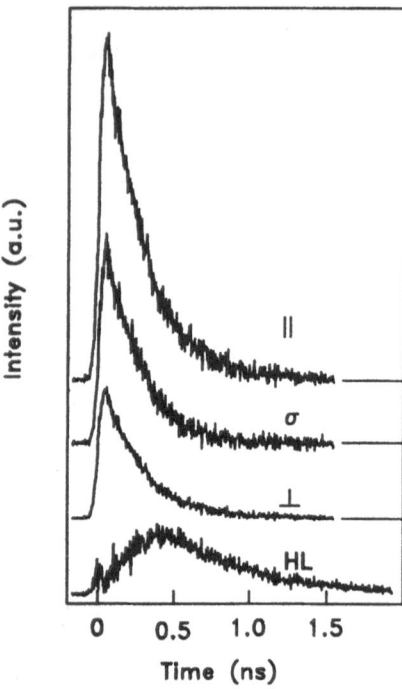   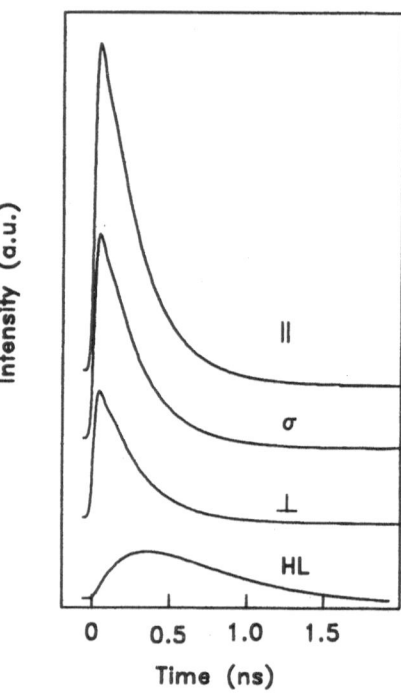

Fig. 6.8 Decomposition of the intensity of the 2TO(L) scattering process in a polarized Raman and an unpolarized hot luminescence component (HL).

Fig. 6.9 Temporal dependence of the polarized Raman and unpolarized hot luminescence contribution calculated with the transform–limited theory using the parameters discussed in the text.

order to be able to apply the theory of Sect. 5.3 we have to identify the rates γ_2, γ_3, which are assumed to be equal, with the rate Γ, and γ_1 with Γ_{ER}. For the fit, the parameters have been chosen according to the following criteria

1. At $B = 0$ they must give the right ratios of the polarized intensities. This requires as a first step the decomposition of the measurements into an unpolarized background and the true polarized part. The result of this decompostion is presented in Fig. 6.8 and compared to the results of the transform–limited theory. For the calculations a total dephasing rate $\Gamma = 4 \cdot 10^9$ s^{-1} and an energy relaxation rate of $\Gamma_{ER} = 2 \cdot 10^9$ s^{-1} have been assumed. These relaxation rates correspond to a dephasing time $\tau_{coh} = 500$ ps and to an energy relaxation time of $T_1 = 500$ ps, so that the pure dephasing and the energy relaxation contribute equal amounts to the coherence time of the exciton state at this exciton kinetic energy.

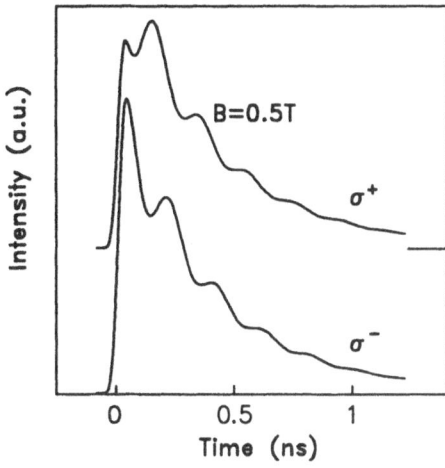

Fig. 6.10
Calculated time dependence of the 2TO(L) scattering line for left (σ^+) and right circularly (σ^+) polarized light. Magnetic field $B = 0.5$ T. For further discussion see text.

2. At $B \neq 0$ the calculations must reproduce the circularly polarized components and their ratio to the unpolarized background. The typical result for such a calculation of the temporal decay is shown in Fig. 6.10. In agreement with the slightly different excitation energy for these measurements (here $\epsilon = 0.5$ meV) the relaxation rates are found to be nearly equal to those at zero field.

From a detailed analysis the following best set of parameters for the transition matrix elements have been found

$$\sigma_S = 1.5 - 1.5i \quad \sigma_T = 0.15 \quad \sigma_T' = 0.075 \ . \tag{6.58}$$

According to Figs. 6.9 and 6.10 these give a very reasonable fit of the experimental data.

These results for the parameters that characterize the transition moments of the indirect absorption show that both valence band and conduction band scattering contribute with nearly equal amounts to σ_S (see Eq. (8.43)). Therefore, because of the larger energy denominator, the strength of the conduction band scattering matrix elements have to be much larger than the valence band scattering matrix elements, a somewhat unexpected behaviour. The absorption into the pure triplet states which can be calculated from these data is expected to be about 1% of the singlet absorption. Here it would be highly interesting to confirm the analysis by a direct measurement of the absorption in high purity single crystals showing a very small inhomogeneous broadening of the exciton transition.

To determine the g factors and the exchange energy, measurements of the temporal dependence of the 2TO(L) intensity as a function of magnetic field are needed. Hereby it is sufficient to look at the oscillatory part of the scattered intensity, given as the difference between the left and right circularly polarized components. Two examples for these quantum beat signals are

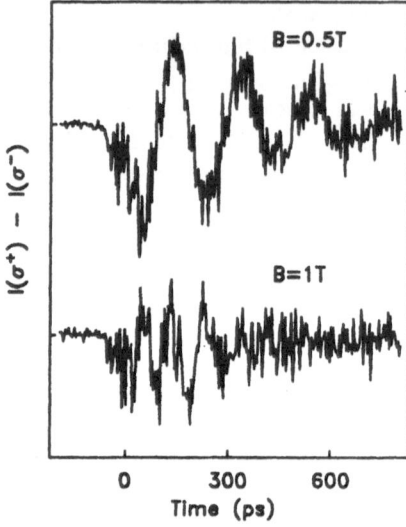

Fig. 6.11 Quantum beat signals in the 2TO(L) scattered intensity at various magnetic fields. Shown is the difference between left and right circularly polarized detected intensity. Excitation conditions as in Fig. 6.7.

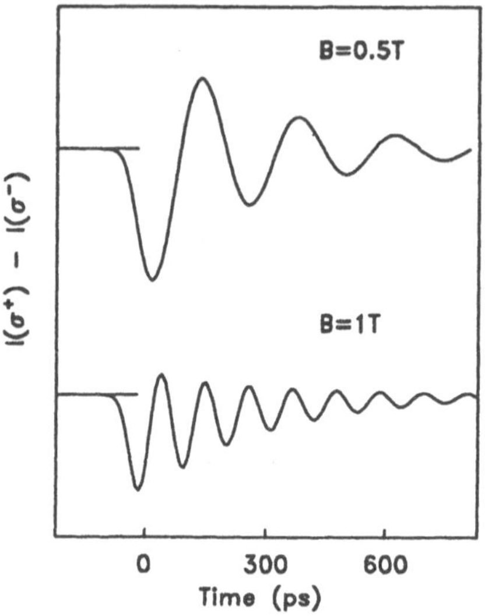

Fig. 6.12 Calculated quantum beat signals of the 2TO(L) process at various magnetic fields.
All conditions as in Fig. 6.11. For further explanations see text.

displayed in Fig. 6.11 [150], obtained at a magnetic field of 0.5 and 1 T. As expected from the increase of the splitting between the singlet–triplet states with field, the oscillation frequency of the beat approximately doubles. The amplitude of the oscillation, however, decreases drastically for the higher field, while only a minor decrease of the coherence time can be deduced from the analysis of the curves. Calculating the quantum beat signals, this behaviour cannot be explained with the previous assumption $g_v^\perp = 0$ [150, 159]. As demonstrated in Fig. 6.12, a value of $g_v^\perp \neq 0$ leading to a mixing of the exciton states and thus reducing the amplitude of the oscillations allows one to reproduce the experimental quantum beat signals. Here the following parameters have been chosen

$$\Delta = 0.15 \text{ meV} \quad g_c = 1.45 \quad g_v^\| = 2.6 \quad g_v^\perp = 1.0 \qquad (6.59)$$

In contrast to the set of parameters obtained previously from the quantum beat measurements, these data agree much better with the set of magneto-optic parameters from high field measurements $\Delta = 0.17$ meV, $g_c = 1.46$ and $g_v^\| = 2.61$, which are much less sensitive to the exact value for $g_v^\perp$ (compare the high field exciton energies displayed in Fig. 6.2 for $g_v^\perp = 0$ and $g_v^\perp = 1$).

6.5 Exciton–Phonon Relaxation Processes

To quantitively investigate the relaxation processes of indirect excitons and their dependence on the exciton energy, time–resolved light scattering has turned out to be the ideal method. In such a measurement, the exciton system is excited via the indirect absorption process by a short laser pulse to produce excitons with a well–defined energy. The temporal evolution of the exciton system can then monitored by the radiative recombination process, which, in the *absorption followed by emission* approximation, reflects the population of the exciton states. The various relaxation channels possible for the exciton, e.g. by phonons, then show up as sidebands of the 2TO(L) process shifted by the loss in excitonic energy induced by the scattering event.

As previous investigations of resonant light scattering under stationary conditions have shown (for a recent review see [42]), besides the capture of excitons by impurities and defects, which dominates the exciton relaxation at small kinetic energies, in the silver halides three different phonon–induced relaxation processes are important.

1. Scattering by long–wavelength acoustic LA phonons inside each of the four L point valleys (intravalley scattering)

2. Scattering by X phonons between inequivalent L valleys (intervalley scattering)

3. Intravalley scattering via polar optical phonons.

In the first time–resolved studies [22], values for the excitonic relaxation rates for the LA(Γ) and X phonon scattering were obtained, at this time, however, analyzing the experiments in the framework of the rate–approximation [40, 22].

In this work we have demonstrated the importance of the coherent contributions to the scattering processes demanding an analysis of the experimental results in the framework of the transform–limited theory by taking the excitation and detection process fully into account. As any coherent contribution to the scattering process shows up as a non–zero degree of polarization of the scattered intensity, the way to characterize the influence of the coherence of the exciton states is straightforward. As examples of this procedure, the TA(X) and the LO(Γ) exciton phonon scattering processes will be considered. At low temperature, these occur for excitation photon energies larger by the amount of the corresponding phonon energy than the indirect absorption edge as a 2TO(L) + TA(X) and 2TO(L) + LO(Γ) scattering line, respectively, in the spectra.

In Fig. 6.13 the time dependence of the polarized intensity of the 2TO(L)+ TA(X) process is compared to that of the 2TO(L) scattering excited at the same photon energy. Due to the higher exciton kinetic energy compared to the measurements shown in the last section, the decay time of the 2TO(L)

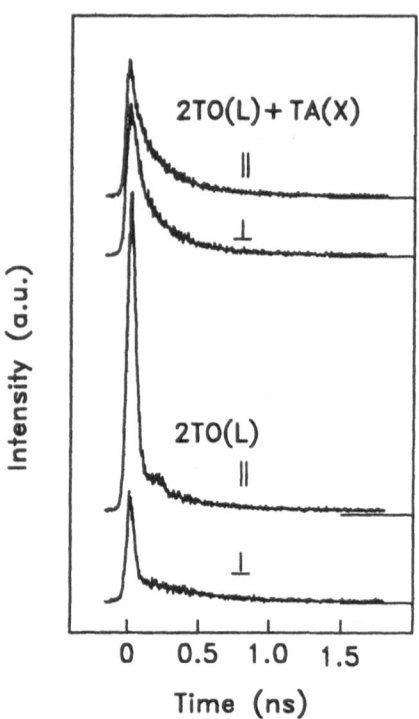

Fig. 6.13
Temporal behaviour of the 2TO(L) + TA(X) and the 2TO(L) scattering process for different polarization directions of the scattered light. Excitation at $E_L = 2.7027$ eV. The horizontal lines on the right–hand side mark the zero of the intensity for each curve.

scattering line is much shorter. Furthermore, a decompostion of the scattered light into the Raman and hot luminescence contributions shows that the hot luminescence background is much weaker than at smaller exciton kinetic energies and the 2TO(L) line shows the maximum possible degree of polarization of $P = 0.5$. This implies that in this energy range pure dephasing processes of the exciton are negligible. In contrast, the 2TO(L)+TA(X) scattering line is completely unpolarized under the excitation conditions of the experiment. However, this observation alone does not allow one to conclude that the coherence of the exciton is destroyed during the scattering with the TA(X) phonons since, due to the degeneracy of the phonon mode, a cancellation of the polarization is possible. Here further, more detailed measurements, in particular the investigation of quantum beats of this scattering process, are necessary.

Because of its simpler symmetry, for the excitonic relaxation process involving LO(Γ) phonon scattering the polarized measurements allow a definite conclusion concerning the phase relaxation of the excitonic state. The temporal dependence of the corresponding two scattering lines, the 2TO(L) and the 2TO(L) + LO(Γ) process, are displayed in Fig. 6.14. Here the excitation photon energy was chosen slightly above the threshold for the onset of LO(Γ) scattering at $\epsilon = \hbar\Omega_{LO(\Gamma)} = 17.0$ meV ([42]) corresponding to an exciton kinetic energy of $\epsilon = 17.1$ meV. As can be seen from the decay curves, at this

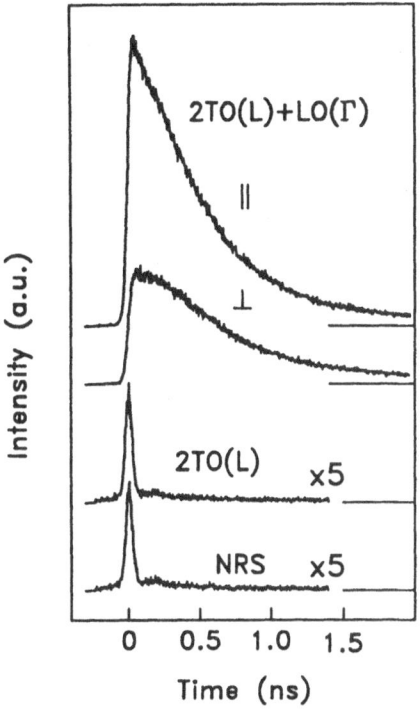

Fig. 6.14
Temporal behaviour of the 2TO(L) + LO(Γ) and the 2TO(L) scattering processes for different detection polarization directions. Excitation at E_L = 2.7097 eV. The lines on the right–hand side mark the zero of the intensity for each curve.

excitation energy the decay of the primary 2TO(L) scattering process has become so short that it could not be resolved within the experimental time resolution, which for these measurements amounted to about 40 ps. Comparing the time behaviour with that of the non–resonant Raman background measured at a Raman shift lower by 1 meV as $2\hbar\Omega_{TO(L)}$ (lowest curve in Fig. 6.14) shows that the time curve of the 2TO(L) line directly follows that of the laser pulse. But still there is a difference in the peak height of the two curves as shown by the difference curve in Fig. 6.15.

The temporal behaviour ot the intensity of the 2TO(L) + LO(Γ) scattering process agrees fairly well with that of the 2TO(L) process excited with photon energy lower by one LO phonon energy. This can be seen by comparing the decomposition of the polarized intensity of this scattering process into the Raman and hot luminescence components shown in Fig. 6.15 with the corresponding curves of Fig. 6.8. This suggests that the coherency of the exciton states is conserved during the scattering with LO(Γ) phonons. To substantiate this conclusion a direct observation of the quantum beating in the intensity of the 2TO(L) + LO(Γ) scattering process would be highly desirable.

The obvious difference between the behaviour of the LO(Γ) and TA(X) scattering is related to the fact that the LO(Γ) phonon is a single, non-degenerate mode, while the TA(X) phonon is twofold degenerate and the modes mix the exciton states in the scattering process.

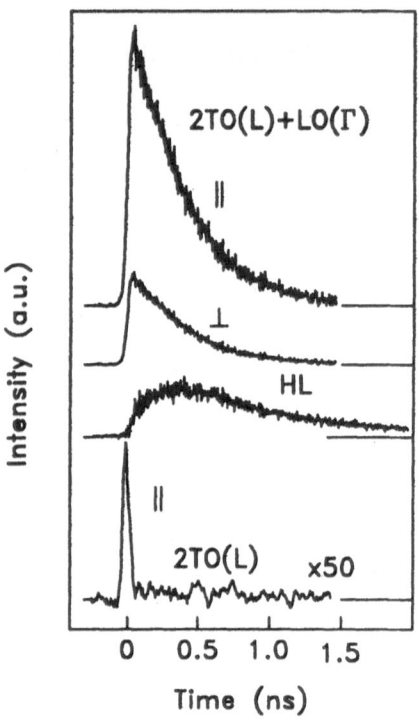

Fig. 6.15
Decomposition of the temporal behaviour of the $2TO(L) + LO(\Gamma)$
and the $2TO(L)$ scattering line after picosecond excitation, taken from
Fig. 6.14, into Raman and hot luminescence contributions with different
detection polarization directions. The
lines on the right–hand side mark the
zero of intensity for each curve.

As the temporal decay of the Raman component of the primary $2TO(L)$
scattering line is determined by the total relaxation rate of the resonantly
excited exciton state, a measurement of the decay of the $2TO(L)$ line as a
function of excitation photon energy allows one to explore the dependence
of the excitonic relaxation rate on the exciton kinetic energy. At high kinetic
energies, the contribution of pure phase relaxation processes to the decay rate
is negligible, in other words, here we measure directly the excitonic energy
relaxation rate. For this purpose a series of measurements of the decay of the
$2TO(L)$ line was analyzed in the framework of the transform–limited theory
of Sect. 5.3. Thereby the difference between the intensity polarized parallel
and perpendicular to that of the exciting laser pulse was modeled assuming for the laser pulse and the spectral filter single–sided exponentials. The
theoretical curves were finally convoluted with the pulse broadening function of the photomultiplier, which was measured separately. In this way, even
extremely short relaxation times could reliably be obtained from the measurements. This analysis gives the coherence time, i.e. the homogeneous linewidth
of the exciton states. The results are displayed as open diamonds in Fig. 6.16.
For exciton kinetic energies higher than the onset of $LO(\Gamma)$ scattering the
analysis of the time curves was not able to give reliable results for the decay
time which become too short compared to the time resolution of our system.
Comparing the time–integrated intensities of the $(2TO(L) + LO(\Gamma))$ and the

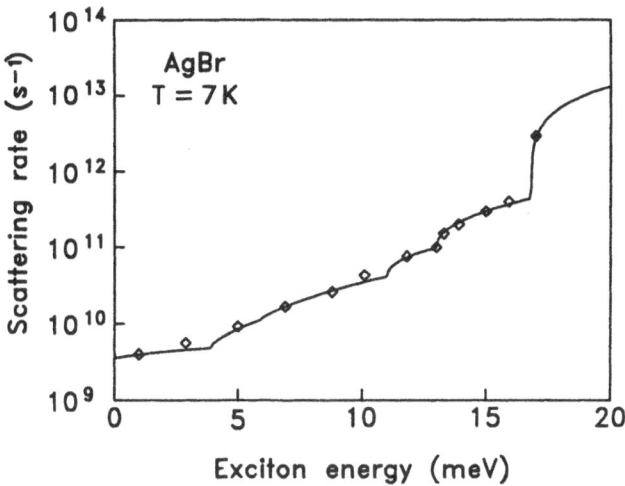

Fig. 6.16 Energy dependence of the total relaxation rate of the indirect exciton in AgBr. Full line: calculation; open diamonds: data obtained from the analysis of the decay of the intensity of the 2TO(L) line; full diamond: value obtained from the intensity ratio as discussed in the text.

2TO(L) scattering processes, however, allows us to determine indirectly the LO(Γ) scattering rate. Considering the Raman contribution to the scattering process in both cases, the time–integrated intensities can be obtained in a simple cascade model giving

$$\frac{\Gamma(\epsilon - \hbar\Omega_{LO})}{\Gamma(\epsilon)} = \frac{I_{2TO(L)+LO(\Gamma)}}{I_{2TO(L)}} \tag{6.60}$$

whereby $\Gamma(\epsilon)$ denotes the total (coherent) excitonic decay rate. Using this equation, from the measurements a decay rate of $\Gamma(\epsilon = 17.1 \text{ meV}) = 3(\pm 2) \cdot 10^{12}\text{s}^{-1}$ can be deduced corresponding to a coherent exciton lifetime of about 300 fs.

To quantitatively understand the energy dependence of the experimental data the various exciton–phonon scattering processes of the indirect exciton have to be considered. From the previous investigations, which have been restricted to the low kinetic energy part of the diagram, it is well known that the important relaxation channels at small kinetic energies are scattering by long–wavelength acoustic (LA(Γ)) phonons and intervalley (TA(X) and LA(X)) phonons. Also it is suggested by stationary investigations and theoretical calculations that at high kinetic energies the scattering by LO(Γ) phonons is the dominant process [42].

From the expressions given in Sect. 4.4 for these relaxation mechanisms, the total excitonic scattering rate is calculated as the dashed curve in Fig. 6.16. Here for the LA(Γ) and intervalley scattering by TA(X) and LA(X) phonons the rate constants

$$c_{ac} = 1.85 \cdot 10^{10}\text{s}^{-1}\text{meV}^{-3}, \quad c_{TA(X)} = 5.5 \cdot 10^{8}\text{s}^{-1}\text{meV}^{-3/2},$$
$$c_{LA(X)} = 2.5 \cdot 10^{8}\text{s}^{-1}\text{meV}^{-3/2}$$

given in Ref. [22] were used. The rate constant for the LO scattering by the Fröhlich interaction was calculated from expression (4.79) using appropriate physical constants from [158]. Somewhat uncertain in this calculation is the value of the hole mass to be inserted in the expression, since it is highly anisotropic. The best compromise seems to be the density of state mass $\overline{m}_h = (m_h^{\parallel}(m_h^{\perp})^2)^{1/3}$ as it gives the density of final states most important in the derivation of the scattering rate. Besides these phonon relaxation processes, a constant contribution $\Gamma_0 = 3 \cdot 10^9$ s^{-1} was added, which takes into account both the capture at impurities and the pure dephasing.

With these assumptions a good agreement between theory and experiment is achieved in the range of small exciton kinetic energies ($\epsilon < 10$ meV). This, however, is to be expected as the relaxation rates have been calculated from parameters that should be sample independent and have been obtained by fitting previous experiments. The only parameter that is changing from sample to sample is the relaxation rate Γ_0, which depends on the impurity content of the sample under investigation.

For exciton energies higher than 11 meV above the band minimum, the calculated rates strongly disagree with the experimental data. This can be explained by the assumption that in this energy range further relaxation channels open up which are not included in the theoretical relaxation model. Looking at the phonon dispersion (see e.g. [161]), possible candidates for such a process are intravalley scattering by TO(Γ) and intervalley scattering by TO(X) phonons. The onset energies for these relaxation processes given by the corresponding phonon energies are at $\hbar\Omega_{TO(\Gamma)} = 11.0$ meV and $\hbar\Omega_{TO(X)} = 13.0$ meV, respectively. The scattering by TO(Γ) phonons is mediated by the optical deformation potential, which, similar to germanium, can be non-zero at the L point of the BZ and interacts therefore only with the hole component of the exciton. The relaxation rate of these process depends in a simple square–root law on the exciton energy (compare Eq. (4.64)). For the TO(X) intervalley scattering a dependence as in the case of the TA(X) scattering given by Eq. (4.80) is obtained.

Including these relaxation processes allows one to fit the experimental points over the whole range of exciton energies. The rate constants of the additional processes are determined from the fit to be

$$c_{TO(\Gamma)} = 3 \cdot 10^{10}\text{s}^{-1}, \quad c_{TO(X)} = 0.7 \cdot 10^{10}\text{s}^{-1}.$$

These rates are larger by one order of magnitude than those of the acoustic intervalley modes.

The quantitative determination of excitonic relaxation rates down to fractions of a picosecond demonstrates the capability of the method of time–resolved light scattering under transform–limited conditions to explore the dynamics of an exciton system.

7. Resonant Light Scattering of Exciton–Polaritons in Cuprous Oxide

In this chapter, measurements will be presented of resonant light scattering from the $1S$ exciton–polariton of the yellow exciton series in Cu_2O. These will provide an example that allows us to check the theoretical concepts developed in Chap. 5 for exciton-polaritons as resonant intermediate states in the light scattering process. In particular, the possibilities to discriminate between Raman scattering and hot luminescence processes will be explored using the polarization properties of the scattered light and the quantum beating occurring in the Raman scattering by splitting the exciton states in a magnetic field [76, 164, 165].

7.1 Properties of Excitons in Cu_2O

7.1.1 Band Structure and Exciton States

Cuprous oxide (Cu_2O) has a very simple structure crystallizing in a simple cubic lattice (scc) with two formula units in the elementary cell. The oxygen atoms thereby form a body-centred- , the copper atoms a face-centred-cubic lattice. As depicted in Fig. 7.1, the Cu_2O has a direct band gap at the Γ point of the Brillouin zone. However, due to the copper d-electrons the ordering of the bands is different from that in other cubic structures, like the III-V compounds as the upper valence bands are formed by the $3d$ electrons and have symmetry $\Gamma_7^+ + \Gamma_8^+$. As the lowest conduction band also has even parity, the optical transition between these band states is dipole forbidden.

The exciton states formed out of the various band states in Cu_2O have been well known for a long time and have actually become model systems for a wealth of investigations (for a review see [166]). Here we are interested mainly in the energetically lowest exciton state, the $1S$ exciton of the *yellow* series that is composed of the Γ_7^+ of the valence and of the Γ_6^+ states of the conduction band. From the symmetry relations

$$\Gamma_{ex,1S} = \Gamma_7^+ \otimes \Gamma_6^+ \otimes \Gamma_1^+ = \Gamma_2^+ + \Gamma_5^+ \tag{7.1}$$

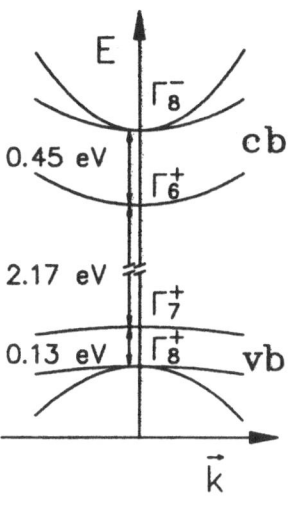

Fig. 7.1
Band structure of Cu$_2$O near the Γ point. cb and
vb: conduction and valence bands. From [76].

follows that optical transitions into the threefold degenerate Γ_5^+ states, also designated as the ortho–exciton, in lowest order are possible by the electrical quadrupolar interaction. The corresponding absorption band is extremely weak with absorption constants of the order of some cm^{-1}. The non-degenerate Γ_2^+ state has pure triplet character and is usually known as the para–exciton. Optical transitions into this state are forbidden in all orders and may become allowed only by an external perturbation.

For purposes of symmetry, the Γ_5^+ states can be taken to be of the form

$$|x\rangle = \Psi_{yz} \quad |y\rangle = \Psi_{xz} \quad |z\rangle = \Psi_{xy} \, , \qquad (7.2)$$

whereby the close analogy to the d–states of the hydrogen atom becomes evident, e.g., in that the function Ψ_{yz} transforms like the product of the cartesian coordinates y and z [107].

Applying an external magnetic field B the exciton states split as shown in Fig. 7.2. This can be described by the effective Hamiltonian

$$H = \frac{1}{2}(g_c + g_v)\mu_B F \cdot B \qquad (7.3)$$

where F denotes a pseudospin vector with z component $F_z = \pm 1, 0$, g_c and g_v are the g factors of valence and conduction band and μ_B is the Bohr magneton (see e.g. [167]). The diagonalization of H gives the splitting energies and the coupling of the basis states (7.2) to the states with $F_z = 0, \pm 1$ depending on the magnetic field. For B along the [001] direction, e.g., the states with $F_z = \pm 1$ are given by the linear combinations $\Psi_{yz} \pm i\Psi_{xz}$ The pseudospin of the para–exciton state is $F = 0$. But due to the large exchange interaction of the $1S$ exciton (splitting between ortho and para state 11.9 meV) the mixing of the para-exciton to the ($F_z = 0$) state of the ortho-excitons can be neglected at small magnetic fields [168].

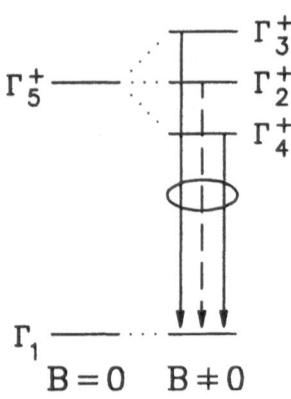

Fig. 7.2
Splitting of the Γ_5^+ exciton states in a magnetic field B.

Due to the cubic symmetry, the dispersion relation of the $1S$ exciton state is of simple parabolic form (Eq. 4.7) with an effective mass $M^* = 2.7m_0$ [40].

7.1.2 The $1S$ Exciton–Polariton

Despite the small oscillator strength connected with the quadrupole transition into the Γ_5^+ states, which amounts to $f_{[001]} = 3.6 \cdot 10^{-9}$ [115], the interaction with light leads to the formation of polaritons as the true eigenstates of the crystal (see Sect. 4.2.3).

The polariton dispersion relations for the Γ_5^+ states at zero magnetic field $B = 0$ are depicted in the left part of Fig. 7.3. In order to make the small splitting between the polariton branches visible, the curves shown in Fig. 4.2 have been expanded in scale. For the calculation the following parameters have been used: $M^* = 2.7m_0$; quadrupole energy $E_Q = \hbar\omega_0 = 2.0329$ eV; dielectric constant $\varepsilon_\infty = 6.5$. The right part of the figure shows the dependence of the polariton group velocity on the energy as calculated from the dispersion relations according to the relation $v_g = d\omega/dk$. The curves show the strong variation in group velocity from the velocity of light in the medium $(c/\sqrt{\varepsilon_\infty})$ down to the exciton group velocity $\sqrt{2M^*\varepsilon}$ (with kinetic energy ε) by changing the polariton energy only by a small amount.

7.1.3 Raman Processes

Besides directly, via the quadrupole interaction (EQ), the $1S$ exciton can also interact with light by phonon–assisted scattering processes similar to an indirect exciton. These transitions proceed via energetically higher band states, the transitions to which are dipole allowed. These are e.g. the Γ_8^- conduction band states. The phonons participating in the indirect transition have to be of odd parity $(ED(\Gamma_n^-)$ transition). From the selection rules $\Gamma_2^-, \Gamma_3^-, \Gamma_4^-$ and Γ_5^- phonons are possible. Experimentally, the transitions involving Γ_3^-

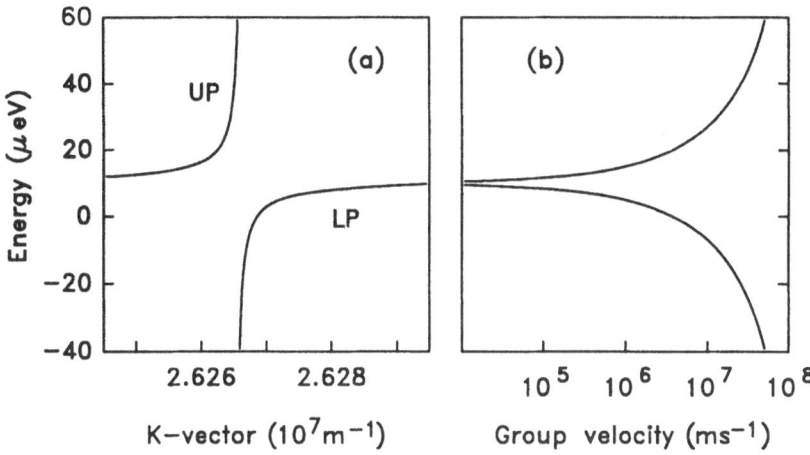

Fig. 7.3 (a) Calculated polariton dispersion $E(k)$ of the $1S$ exciton in Cu_2O. Zero of the energy scale is at the quadrupole resonance E_Q. (b) Dependence of the group velocity on the polariton wavevector. After [76].

phonons are the most intense [39]. Combining the quadrupole transition with the phonon-assisted process, one–phonon Raman scattering results which is resonant with the $1S$ state either with the exciting photon $(EQ - ED(\Gamma_3^-)$ process, *incoming resonance*) or with the scattered photon $(ED(\Gamma_3^-) - EQ$ process, *outgoing resonance*).

In ultrapure samples, like the one at our disposal[1], the one–phonon Raman lines dominate the light scattering spectrum as is shown in Fig. 7.4 for an excitation photon energy coinciding with the quadrupole resonance. The two strongest lines are ascribed to scattering with a Γ_3^- phonon at 13.6 meV and a Γ_2^- phonon at 10.5 meV. It is interesting to note that the spectral linewidth of the Raman lines is much smaller than that of the exciting laser indicating a highly resonant scattering cross section. At the high energy side of each scattering line a weak sideband occurs due to the emission of thermalized excitons.

In addition to one–phonon scattering, similar to the case of AgBr, two–phonon processes become resonantly enhanced at the $1S$ exciton. They involve the phonon–assisted dipole interaction in both the absorption and the emission process and they start one phonon energy above the $1S$ quadrupole transition.

In the past, the various resonance Raman processes occurring in Cu_2O have been investigated thoroughly, whereby both stationary [39, 40] as well as time-resolved measurements [46] have been performed. One of the main topics of these studies was to discriminate experimentally between the true Raman scattering and hot luminescence processes that occur at the same

---

[1]This sample was kindly supplied by D. Fröhlich, University of Dortmund.

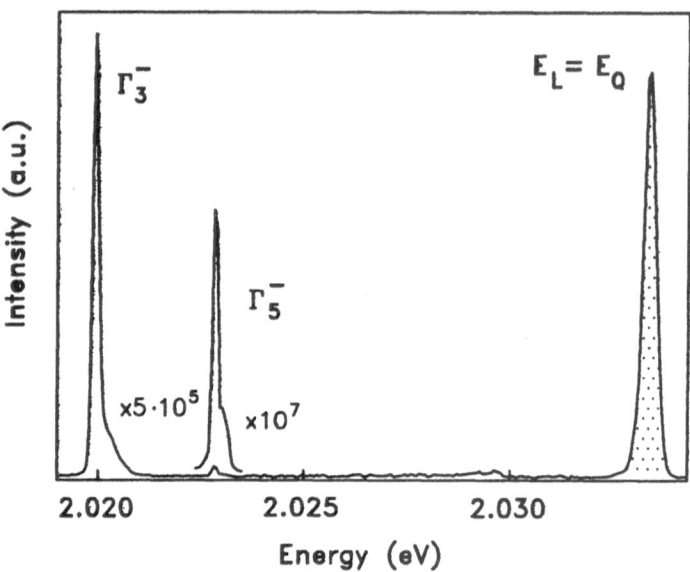

Fig. 7.4 Resonance Raman spectrum of the Cu_2O sample in transmission geometry. The excitation photon energy E_L (laser line marked by dotted area) is resonant with the quadrupole transition at $E_Q = 2.0329$ eV with wavevector $k_L \| [001]$ and polarization $e_L \| [100]$. $T = 2$ K. From [76].

energy shift from the laser line. Here we will show, how the use of quantum beat spectroscopy allows one to resolve this old and still open question.

In all these studies, however, the polariton properties of the $1S$ exciton have been neglected completely, as it was argued that they are not important because of the extremely low oscillator strength of the transition. In contrast, recent experiments in our group that will be discussed in the following sections have demonstrated the importance of the polariton character in the quantitative interpretation of the time-resolved light scattering experiments [76, 164]. In this respect, the results of the Raman experiments correspond to recent studies of time-resolved resonant optical transmission in Cu_2O which showed the importance of the polariton as a coherently propagating wave even at small oscillator strengths [115, 169, 170].

7.2 Relaxation Dynamics

Before discussing the experimental results it is useful to consider the optical selection rules and the relaxation dynamics of the Γ_5^+ exciton states in detail, since due to the directional dependence of the quadrupole interaction some peculiarities do occur.

7.2.1 Density Matrix Formalism

First, the relaxation dynamics will be treated in the density matrix formalism. To this end, the exciton density matrix has to be decomposed into irreducible components. The Γ_5^+ symmetry of the exciton states leads to the following decomposition of ρ

$$\Gamma_5^+ \otimes \Gamma_5^{+*} = \Gamma_1^+ + \Gamma_3^+ + \Gamma_4^+ + \Gamma_5^+ . \tag{7.4}$$

They correspond to the following linear combinations of density matrix elements

$$\rho^{(1)} = \frac{1}{\sqrt{3}}(\rho_{xx} + \rho_{yy} + \rho_{zz}) \tag{7.5a}$$

$$\rho^{(3)} = \frac{1}{\sqrt{6}} \left(\begin{array}{c} -\rho_{xx} - \rho_{yy} + 2\rho_{zz} \\ \sqrt{3}(\rho_{xx} - \rho_{yy}) \end{array} \right) \tag{7.5b}$$

$$\rho^{(4)} = \frac{1}{\sqrt{2}} \left(\begin{array}{c} \rho_{yz} - \rho_{zy} \\ \rho_{zx} - \rho_{xz} \\ \rho_{xy} - \rho_{yx} \end{array} \right) \tag{7.5c}$$

$$\rho^{(5)} = \frac{1}{\sqrt{2}} \left(\begin{array}{c} \rho_{yz} + \rho_{zy} \\ \rho_{zx} + \rho_{xz} \\ \rho_{xx} + \rho_{xx} \end{array} \right) . \tag{7.5d}$$

Inserting these relations into the density matrix results in

$$\rho = \tag{7.6}$$

$$\left(\begin{array}{ccc} \frac{1}{\sqrt{6}}(\sqrt{2}\rho^{(1)} - \rho_1^{(3)} + \sqrt{3}\rho_2^{(3)}) & \frac{1}{\sqrt{2}}(\rho_3^{(5)} + \rho_3^{(4)}) & \frac{1}{\sqrt{2}}(\rho_2^{(5)} - \rho_2^{(4)}) \\ \frac{1}{\sqrt{2}}(\rho_3^{(5)} - \rho_3^{(4)}) & \frac{1}{\sqrt{6}}(\sqrt{2}\rho^{(1)} - \rho_1^{(3)} - \sqrt{3}\rho_2^{(3)}) & \frac{1}{\sqrt{2}}(\rho_1^{(5)} + \rho_1^{(4)}) \\ \frac{1}{\sqrt{2}}(\rho_2^{(5)} + \rho_2^{(4)}) & \frac{1}{\sqrt{2}}(\rho_1^{(5)} - \rho_1^{(4)}) & \frac{1}{\sqrt{6}}(\sqrt{2}\rho^{(1)} + 2\rho_1^{(3)}) \end{array} \right) .$$

Here $\rho_l^{(n)}$ denotes the l–th component of the irreducible density matrix that belongs to the representation Γ_n.

The relaxation of each of these irreducible components is given by one relaxation rate constant that is denoted as γ_i with $i = 1, 3, 4, 5$. Using these relations, the time dependence of the density matrix can be calculated for all cases.

7.2.2 Optical Coupling Matrix Elements

Here we are interested in the coupling of the Γ_5^+ exciton states with light by the quadrupole and the Γ_3^- phonon-assisted dipole interaction. The corresponding coupling matrix elements can be calculated by a group theoretical analysis using the tables in Ref. [100] assuming the 0° scattering geometry

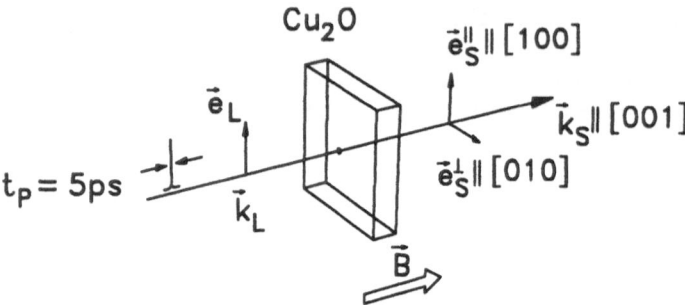

Fig. 7.5 Transmission scattering geometry used in the Raman experiments in
Cu$_2$O. B: magnetic field; k_L, k_S and e_L, e_S: wavevectors and polarization directions
of exciting laser and scattered light. After [76].

shown in Fig. 7.5. In this transmission geometry, the polarization basis vectors can be taken to be the same for both the incident and the scattered light,
so that one set of coupling matrices is sufficient. For the quadrupole interaction, which depends on the direction of the wavevector of light, the two cases
$k\|[001]$ (case A) and $k\|[110]$ (case B) are considered. The polarization basis
vectors in case A are given by

$$e_x = [100] \quad e_y = [010] \tag{7.7}$$

while in case B they are

$$e_x = [\bar{1}10] \quad e_y = [001] . \tag{7.8}$$

The exciton states are given by (7.2). Then for the quadrupole coupling matrices the following results are obtained

$$\text{case A}: \ \underline{\underline{M}}^Q = \begin{pmatrix} 0 & 1 & 0 \\ 1 & 0 & 0 \end{pmatrix} \tag{7.9}$$

$$\text{case B}: \ \underline{\underline{M}}^Q = \frac{1}{\sqrt{2}}\begin{pmatrix} 0 & 0 & 0 \\ 1 & 1 & 0 \end{pmatrix} . \tag{7.10}$$

In the case of the Γ_3^- phonon-assisted dipole coupling the twofold degeneracy
of the phonon modes has to be taken into account. Therefore, two sets of
coupling matrices exist and the results for the scattered intensity have to be
added incoherently for the two modes. The matrices are as follows

$$\text{case A}: \ \underline{\underline{M}}_1^D = \frac{\sqrt{3}}{2}\begin{pmatrix} -1 & 0 & 0 \\ 0 & 1 & 0 \end{pmatrix} \tag{7.11}$$

$$\underline{\underline{M}}_2^D = -\frac{1}{2}\begin{pmatrix} 1 & 0 & 0 \\ 0 & 1 & 0 \end{pmatrix} \tag{7.12}$$

$$\text{case B}: \quad \underline{\underline{M}}^D_1 = \sqrt{\frac{3}{8}} \begin{pmatrix} 1 & 1 & 0 \\ 0 & 0 & 0 \end{pmatrix} \tag{7.13}$$

$$\underline{\underline{M}}^D_2 = \sqrt{\frac{1}{8}} \begin{pmatrix} 1 & -1 & 0 \\ 0 & 0 & 1 \end{pmatrix}. \tag{7.14}$$

With these relations all possible polarization effects and quantum beat signals can be calculated for both the one– and two–phonon scattering processes.

7.3 One–Phonon Scattering

7.3.1 Experimental Results

Looking at the coupling matrices of the foregoing section, one immediately notices that the Raman contributions to the one– and two–phonon processes are strongly polarized in agreement with the usual Raman selection rules [171]. The two–phonon Raman process is always found to be completely polarized parallel to the polarization direction of the exciting laser, whereas the one–phonon process shows a directionally dependent polarization anisotropy shown for some cases in Table 7.1. The phase relaxed hot luminescence turns out to be completely unpolarized in all cases, as expected. The four rate constants that govern the decay of the scattered light can be obtained from the analysis of differently polarized time curves. Here we will not go into detail as this requires somewhat tedious expressions. Most important for the further discussion, however, is that for the cases listed in Table 7.1 only the two constants γ_1 and γ_3 occur in the decay, the first one describing the depopulation of the exciton states corresponding to T_1, the other describing the cross relaxation between the degenerate exciton states.

For discrimination of Raman and hot luminescence processes, the table also shows that the best choice is that of case A and excitation with light polarized parallel to the [100] direction, as in this case the parallel polarized intensity component does not contain any Raman contribution at all. Also, in this scattering geometry, pronounced quantum beat signals are expected,

Table 7.1 Raman intensities of the $EQ - ED(\Gamma_3^-)$ process.

k_L	e_L	e_S	I
[001]	[100]	[100]	0
		[010]	1
[001]	[110]	[110]	0.25
		[1$\bar{1}$0]	0.75

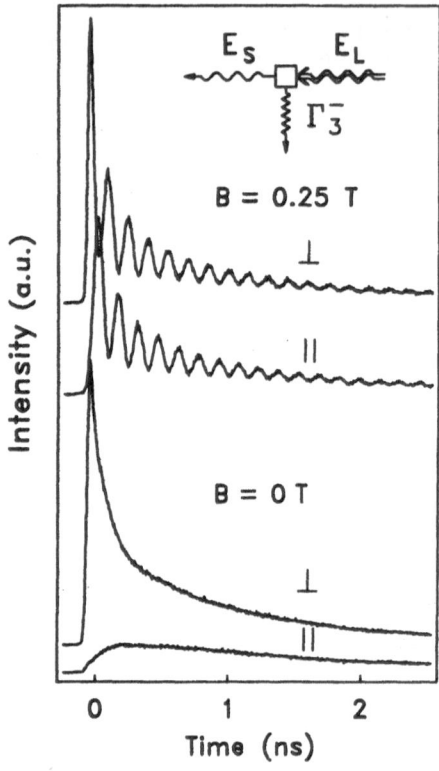

Fig. 7.6
Time dependence of the polarized intensity of the Γ_3^- phonon scattering process for $B = 0$ (lower curves) and $B = 0.25\,\text{T}$ along [001] (upper curves) in Cu_2O at $T = 2.1\,\text{K}$ for resonant excitation at $E_L = E_Q$ as indicated schematically in the figure. The excitation pulse with a duration of 5 ps propagates with $k_L\|[001]$ and is polarized along $e_L\|[100]$. The scattered light with $k_S\|[001]$ is detected with polarization along [100] ($\|$) and [010] ($\perp$). Inset: schematic polariton dispersion with upper (UP) and lower (LP) branch. The scattering by a Γ_3^- phonon is indicated by an arrow. E_Q: energy of the quadrupole transition.

oscillating with frequency $\Delta\nu = |g_c + g_v|\mu_B B/h$. The experimental time dependences of the scattered intensity in various magnetic fields excited at the *incoming resonance* ($E_L = E_Q$) are depicted in Fig. 7.6 for this scattering geometry. As expected, at zero field ($B = 0$) the light scattering is almost completely polarized. This means that Raman processes that proceed under conservation of the coherence of the exciton states are the dominant scattering mechanism. Hot luminescence only gives a small contribution and, therefore, elastic scattering processes giving rise to pure dephasing are of minor importance.

The coherent nature of the intermediate exciton-polariton states can now be demonstrated by the quantum beating of the scattered intensity that occurs on applying a magnetic field (upper curves in Fig. 7.6 at $B = 0.25\,\text{T}$). The oscillations extend over almost the whole time range where the scattered intensity is non-zero. This also indicates that the phase relaxation is dominated by depopulation processes. By directly Fourier transforming the beat signals [76], rather long coherence times of the order of $\tau_{\text{coh}} \simeq 1$ ns can be derived.

Comparing the time profiles of the scattered intensities with and without magnetic field one sees that the quantum beating occurs in both parallel and

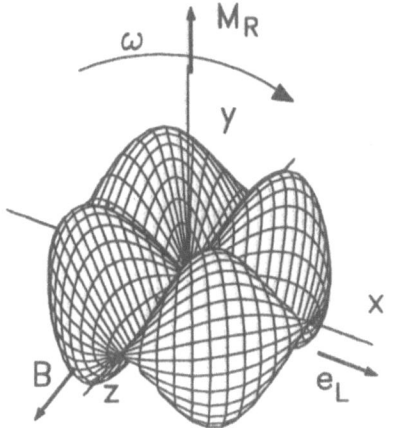

Fig. 7.7
Graphical representation of the Ψ_{xz} state showing the excitation by the quadrupole interaction for light polarized with e_L along x, the Raman transition moment M^D along y and the rotation of the orbital around the magnetic field along z. The scattered light is detected with polarization e_S making an angle Φ to the x axis as indicated.

perpendicular polarization components and in both cases is superimposed on the hot luminescence background which itself is not changed by the field. This directly confirms the result of the transform-limited theory of light scattering that quantum beats show up only in the Raman part of the scattering process that here is given by the difference $I_\parallel - I_\perp$. In contrast, the hot luminescence does not show any beating. The experiments presented here therefore nicely demonstrate the possibility of quantum beating to discriminate experimentally between coherent Raman processes and the incoherent hot luminescence both contributing to the light scattering. At the same time, the continuous decay of the beat amplitude reflects the gradual transformation of the exciton ensemble from the initial coherent state into the phase relaxed statistical mixture.

7.3.2 Quantum Beats and the Time–Resolved Hanle Effect

Despite the quite complex character of the quadrupole interaction, the analogy of the Γ_5^+ exciton states with the d–like atomic orbitals allows a qualitative understanding of the features of quantum beating which turns out to be related to the Hanle effect in the same way as in atomic physics (see [172]).

Under the experimental conditions of the measurements shown in Fig. 7.6 it follows from the coupling matrices that by the quadrupole interaction only the state Ψ_{yz} is excited, the orbital representation of which is displayed in Fig. 7.7. The Raman transition dipole M^D of this state for the Γ_3^- phonon is oriented in the y direction. Applying a magnetic field along z the Ψ_{yz} state can be excited by a short laser pulse as a coherent superposition of the split Γ_3^+ and Γ_4^+ states (see Fig.7.2). However, due to the difference in eigenfrequencies $\Delta\nu = |g_c + g_v|\mu_B B/h$, this superposition state starts to rotate clockwise around the magnetic field axis with frequency $\Delta\nu/2$. Inserting a linear polarizer in the observation direction with its transmission direction at an angle

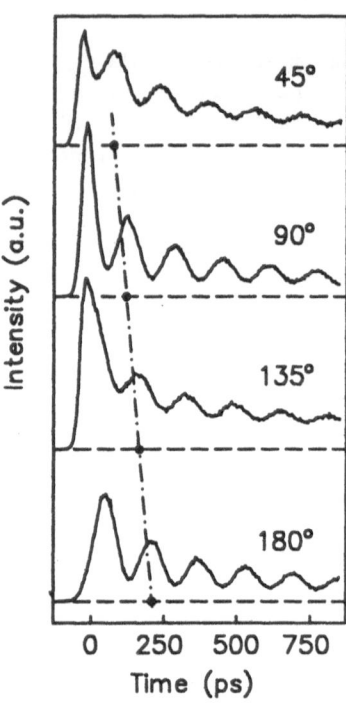

Fig. 7.8
Quantum beating in the time dependence of the Γ_3^- phonon scattering at a magnetic field $B = 0.25\,\mathrm{T}$ applied in the [001] direction and detected at different polarization directions given by the angle Φ measured relative to the $[\overline{1}00]$ axis (see Fig. 7.7). Excitation photon energy $E_L = E_Q$; wavevector $k_L\|[001]$ and polarization of exciting light $e_L\|[100]$. Corresponding beat maxima are marked to illustrate the phase shift of subsequent signals. (After [76]).

Φ to the x axis, it follows immediately that the scattered intensity is modulated at a frequency $\Delta\nu$ thus giving rise to the quantum beats, the maxima occurring when the direction of the Raman dipole is parallel to the polarizer's transmission axis. Therefore, one expects that by changing the angle Φ from π to zero, the maxima in the quantum beat signal shift accordingly to longer times. The rotation of the wavefunction is demonstrated by the measurements shown in Fig. 7.8, thus proving nicely the analogy of the magneto-quantum beating to the Hanle effect also in this case of an exciton–polariton state.

The application of the quantum beat method to the one–phonon scattering process allowed several features of the $1S$ exciton to be studied in some detail, like the precise determination of the g factors of electron and hole from the magnetic field dependence of the oscillation frequency [164]. An interesting application is also the investigation of elastic deformation of the sample through the analysis of a strain induced zero-field splitting [76].

7.3.3 Transform–Limited Model Calculation

Looking more closely at the time dependence of the scattered intensity of the Γ_3^- one–phonon process in Fig. 7.6 at zero field $B = 0$, the parallel polarized component exhibits a rather fast decay at early times, which is visible even better in the difference of intensities $I_\perp - I_\|$ (see upper curve in

Fig. 7.9). This clearly non-exponential decay behaviour cannot be explained in the simple three-level scheme that was implicitly assumed up to now. Rather, this seems to be an effect of the main difference between exciton-polaritons and a simple multi-level system, the propagating nature of the polariton states being expressed by their dispersion relations.

Indeed, by exciting polaritons with a short light pulse that necessarily has to have a finite spectral width all states on the dispersion curves (compare Fig. 7.3) are occupied that have a finite coupling strength with the light field. This coupling is expressed quantum mechanically by the excitonic coupling factors $\tilde{\Pi}_{ex}^l(\boldsymbol{K}, \lambda)$ introduced in Sect. 4.2.3. As can be seen from Fig. 4.2, these coefficients are different from zero in a wide energy range, which, in case of the quadrupole exciton in Cu_2O, amounts to about 0.2 meV around E_Q. The polariton states in this range that are excited by a short light pulse of sufficient spectral width propagate in the sample with very different group velocities v_g. Those with large v_g travel fast through the sample and can therefore interact with phonons only at early times near the time of excitation. However, those with small values of v_g, which correspond to states near the crossing point of the bare photon and exciton dispersions, the so–called *bottleneck region* near the energy E_Q, give rise also to light scattering processes at much longer times. For excitation with spectrally very narrow pulses, however, only polaritons close to E_Q are excited, which differ only a little in their group velocities. Under thess conditions, the non-exponential behaviour should not occur. This was directly verified by comparing measurements with spectrally broad and narrow pulses as shown in the left part of Fig. 7.9.

To calculate the time dependence of the scattered intensity quantitatively, the transform-limited theory of Sect. 5.3 was extended to take the dispersion of the polariton states into account. Thereby the scattering contributions of the triple degenerate polariton states from each branch at the same wavevector have to be added coherently. In contrast, for the one–phonon processes the scattered intensities from states at different wavevector and from the upper and lower polariton branch add up incoherently, as these scattering pathways differ in the final phonon state and therefore are not indistinguishable.

In the case of the *incoming resonance* at $E_L = E_Q$ one obtains the following expression for the scattered intensity of the Γ_3^- Raman process

$$I(t, e_L, e_S) \propto \sum_{k_{u,l}} \sum_{l=1,2} \left| \sum_m g_m^{EQ}(k_{u,l}, e_L) g_{l,m}^{ED}(k_{u,l}, e_S)^* \Psi_{m,k_{u,l}}(t) \right|^2 . \quad (7.15)$$

Here l enumerates the two degenerate Γ_3^- phonon modes and m the Γ_5^+ exciton states. The outer summation is over the wavevectors of the upper and lower polariton branch. g_m^{EQ}, $g_{l,m}^{ED}$ denote the quadrupole and phonon-assisted dipole coupling factors. These can be obtained from the corresponding transition matrix elements and the polariton coupling factors to the photon and

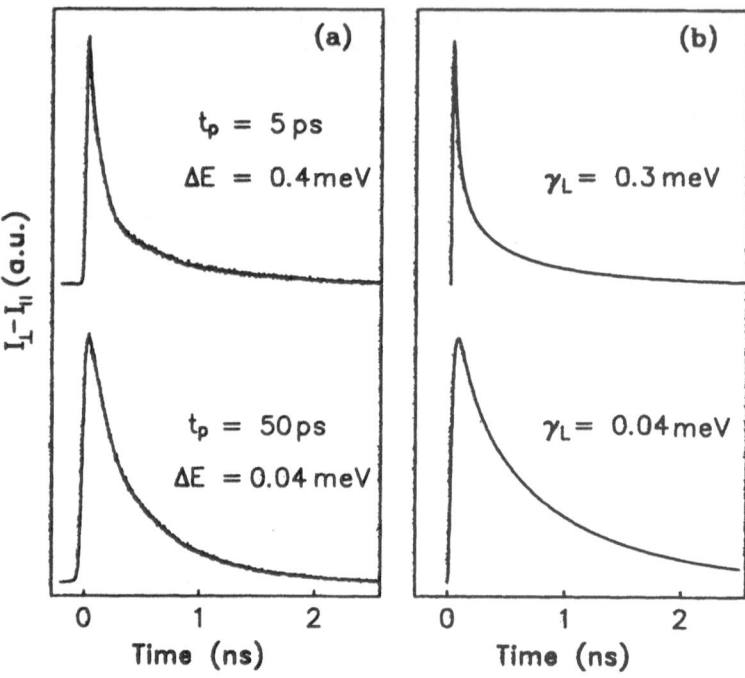

Fig. 7.9 Time dependence of the measured (a) and of the calculated (b) contribution of the coherent part $I_\perp - I_\parallel$ of the Γ_3^- phonon scattering process for excitation with laser pulses differing in spectral bandwidth. ΔE denotes the spectral and t_p the temporal halfwidth of the laser pulses used, γ_L is the spectral width of the laser pulses of 'single–sided exponential' form used in the model calculations.

exciton states as $(\alpha = up, lp)$

$$g_m^{EQ}(K, e_L) = M_m^{EQ} \cdot e_L \tilde{\Pi}_0^\alpha(K, e_L) \qquad (7.16)$$

$$g_{l,m}^{ED}(K, e_S) = M_{l,m}^{ED} \cdot e_S \tilde{\Pi}_m^\alpha(K, e_S) . \qquad (7.17)$$

Here $M_{l,m}^{ED}$ and M_m^{EQ} are the effective matrix elements (see Sect. 7.2.2). In deriving these relations, we assumed that the transition from the outer vacuum space to the dielectric material simply consists in changing the vacuum ground state of the exciton-polariton field. The additional boundary conditions (ABC) that are normally used to transform the photons into the polariton modes [173] are in this description not necessary any more.

The other quantity that is important for the cross section is the damping of the polariton states given by their homogeneous linewidth which enters via the scattering amplitude $\Psi_{m,k}(t)$. Here the depopulation relaxation processes, which transfer polaritons to other states not contributing to the light scattering, and pure phase relaxation have to be taken into account. These relaxation channels are assumed to contribute a constant amount Γ_0 to Γ.

As a second contribution to the polariton lifetime the finite time-of-flight of the polaritons through the sample has to be added. Neglecting the propagation time of the scattered photons at energy E_S in the sample being of the order of some picoseconds, this damping can be modeled like in atomic beam spectroscopy as

$$\Gamma_{tf}(\boldsymbol{k}) = v_g(\boldsymbol{k})/d \;, \tag{7.18}$$

whereby $v_g(\boldsymbol{k})$ denotes the group velocity of the polariton state at wavevector $\boldsymbol{k}$. A similar relation was proposed recently by Aaviksoo [38] to describe polariton damping in anthracene crystals, whereby he suggests an additional factor $1 - R$ with R the reflectivity. This does not seem to be appropriate in our case, as the perfection of the sample surfaces is not good enough to allow for multiple reflections. Finally, for the homogeneous linewidth Γ the following relation is obtained

$$\Gamma = \Gamma_0 + \Gamma_{tf} \;. \tag{7.19}$$

The lineshape of the excitation pulse was assumed to be of single-sided exponential form (see (5.142)) with a spectral width comparable to that of the experiment. As spectral filter for the detection of the scattered light a Fabry Perot etalon (see (5.143)) was used with a bandwidth of $\hbar\gamma_S = 0.25$ meV.

The results of the calculation are displayed in the right part of Fig. 7.9 and reproduce almost quantitatively the experimental results. It is important to stress that all quantities entering the calculations, like oscillator strength, effective exciton masses and damping constant Γ_0 are know from independent studies [115] and no free parameter is contained in the theory. Only for the shorter laser pulse had the spectral width to be chosen a little bit smaller than the experimental value. This is not really surprising as the experimental pulse with 5 ps duration had a Gaussian lineshape with a much steeper tail than the Lorentzian lineshape used in the model calculations. Finally, one should notice that the good agreement of theoretical calculation and experiment could only be achieved in the ABC-free quantum mechanical coupling of polariton and photon states. The same calculations using the usual ABC's of Pekar [174] are not able to reproduce the observed effects.

Recently, in the same samples time-resolved measurements of light pulses directly transmitted through the sample and resonant in energy with the quadrupole transition were performed [115]. Unlike the Raman intensities, in these experiments the transmitted intensity showed a pronounced temporal modulation even at zero magnetic field. This beating was also explained by the simultaneous excitation of polaritons belonging to the upper and lower branch and to different wavevectors. In contrast to Raman scattering, where quantum mechanical probability amplitudes interfere, and, as was discussed above, no beating occurs at zero field, in the transmission experiment the electric fields of the polariton waves have to be added at the exit plane of the sample resulting in the observed interference pattern.

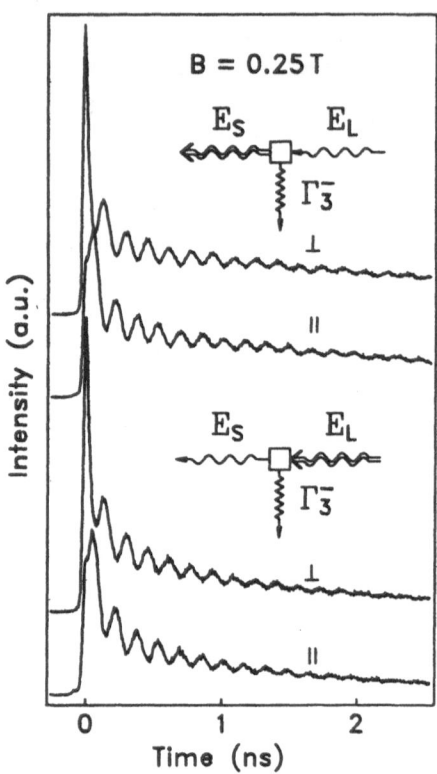

Fig. 7.10

Time dependence of the polarized intensities of the Γ_3^- phonon scattering process at $B = 0.25\,\text{T}$ for excitation in the *outgoing* resonance (upper curves, $E_L = E_Q + \hbar\Omega(\Gamma_3^-)$), and in the *incoming* resonance (lower part, $E_L = E_Q$) with the quadrupole exciton state. Excitation was with light of wavevector $\boldsymbol{k}_L \| [001]$ and polarization $\boldsymbol{e}_L \| [110]$. The scattered light was detected with polarization directions $[110]$ ($\|$) and $[\bar{1}10]$ ($\perp$). The insets depict the scattering process in the outgoing and incoming resonance. The double line denotes the exciton-polariton, the single line the photon outside resonance and the zig–zag line the Γ_3^- phonon.

Exchanging the role of the coupling factors for excitation and detection of the scattered light, Eq. (7.15) also allows one to describe the *outgoing resonance* of the one–phonon scattering process. As the form of the expression is obviously not changed, the intensity and time dependence of the outgoing process should be the same as those of the incoming channel. Experimentally, this prediction can be checked quite easily as is shown by the curves in Fig. 7.10. Here the polarized scattered intensity is compared for the incoming (lower curves) and outgoing (upper curves) resonance. The excitation polarization was chosen here along the [110] direction. In this case, the selection rules (see Table 7.1) predict at zero magnetic field an intensity ratio of 3:1 for the perpendicular and parallel components. As can be seen at early times, where the intensities are not changed from the zero field case, this agrees well with the experiments. Both the intensities and the quantum beat signals are completely identical for incoming and outgoing resonance confirming nicely the assumptions of the theoretical treatment of the scattering. In particular, it follows that indeed the phonons do not influence the coherence of the exciton states and that the summation over the phonon modes has to be performed incoherently.

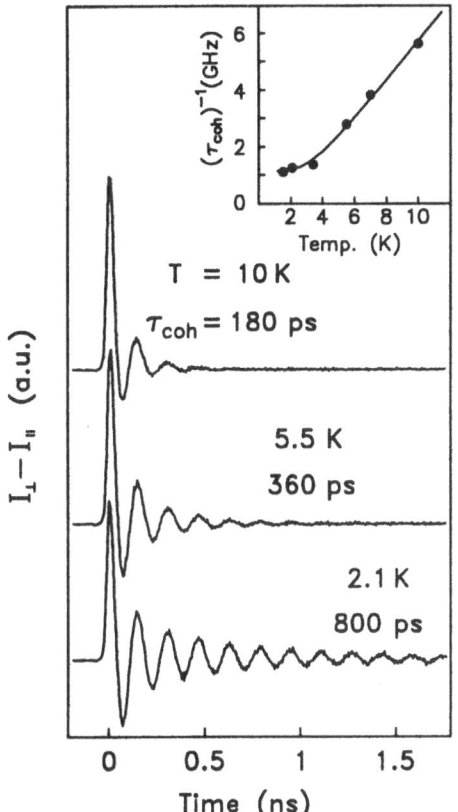

Fig. 7.11
Temperature dependence of the coherent part of the Γ_3^- phonon scattering at a magnetic field $B = 0.25\,T$. Excitation and detection conditions as in Fig. 7.6. Inset: Dependence of the reciprocal of the coherence time τ_{coh} on temperature. The full line is a fit using the polariton relaxation rates as discussed in the text. From [76]).

7.3.4 Temperature Dependence: Polariton–Phonon Scattering

As the last topic of this chapter, the measurement of the temperature dependence of the phase relaxation processes of the exciton-polariton will be discussed. The temperature dependence of the dephasing time τ_{coh} can be deduced by analyzing the difference of the parallel and perpendicular polarized scattered intensities, the quantum beat signal, as depicted in Fig. 7.11 for some temperatures.

Clearly, the decrease of the coherence time with increasing temperature can be seen from the measurements. Quantitative values for τ_{coh} taken from these measurements are shown in the inset. Here the reciprocal of the coherence time, i.e., the homogeneous linewidth is plotted versus temperature.

The linear increase at higher temperatures suggests that scattering by longitudinal acoustic (LA) phonons is responsible for the dephasing. For a quantitative comparison, the scattering rate of the exciton-polaritons in the bottle-neck region was calculated from the relations given in Sect. 4.4 using the following parameters: acoustic deformation potential $D_c - D_v = 2.1$ eV

[175], longitudinal sound velocity $u_L = 4.54 \cdot 10^3 \text{ms}^{-1}$ [176] and density $\rho = 6.1 \cdot 10^3 \text{kgm}^{-3}$ [177]. The dispersion of the polaritons can be neglected in the calculation as it is exciton-like over most of the range of wavevectors. At low temperatures, only emission of phonons is possible and scattering by LA phonons transfers the initial polariton state only into energetically distant final polaritons that have almost completely exciton character. Therefore, in the scattering matrix element only the initial laser excited state has to be replaced by the polariton wavefunction, and to obtain the polariton scattering rates the excitonic scattering rates have to be multiplied by the square of the exciton-polariton coupling factor. In the energy region close to resonance, this multiplication factor amounts to 0.5 as can be seen from Fig. 4.2. The energy relaxation rates calculated by this procedure are shown as the full line in the inset of Fig. 7.11 and by adding a temperature–independent pure phase relaxation rate of $1/T_2' = 0.7 \cdot 10^9 \text{s}^{-1}$ agree perfectly with the experimental results. This means that even at a temperature of 10 K, where the damping is much larger than the splitting of the polariton dispersion curves, optical excitation first creates polariton states that subsequently decay by energy and phase relaxation. This demonstrates nicely the importance of the polariton concept in the treatment of the optical properties of crystals.

8. Time–Resolved Resonant Rayleigh Scattering

In this chapter experimental investigations of time–resolved resonance Rayleigh scattering (RRAYS) from exciton states will be discussed. The experiments are performed on selected III-V and II-VI semiconductor heterostructures and fully confirm the model of this kind of light scattering developed in Chap. 5 from theoretical considerations. In the first section, the relevant exciton states of these compounds are discussed, all crystallizing in the zinc-blende structure. The next section contains the results of stationary measurements of RRAYS in quantum well (QW) structures in the GaAs/AlGaAs system [74, 140] and in thin epitaxial layers of ZnSe/ZnS mixed crystals grown on GaAs substrates [178]. In the third and last section time–resolved investigations in GaAs QW structures are presented. These show for the first time the finite time response of RRAYS [140].

8.1 Excitons in Zinc–Blende–Type Semiconductors

8.1.1 States and Their Dynamics

All semiconductors discussed in this chapter crystallize in the cubic zinc-blende structure with point group T_d. In the bulk, the relevant exciton states are well known [179] and are composed of the lowest conduction band with minimum at the Γ point of the BZ showing only spin degeneracy (symmetry Γ_6) and of the uppermost valence band. Its symmetry Γ_8 corresponds to a $j = 3/2$ fourfold degenerate angular momentum state and is split outside the maximum at $k = 0$ into states with $m_j = \pm 3/2$ (*heavy hole*, hh) and $m_j = \pm 1/2$ (*light hole*, lh) showing a rather complex dispersion ('warping'). The exciton states are in total eightfold degenerate but are split by the exchange interaction according to

$$\Gamma_8 \otimes \Gamma_6 = \Gamma_3 + \Gamma_4 + \Gamma_5 \tag{8.1}$$

into the dipole forbidden states with $J^{ex} = 2$ (Γ_3, Γ_4) and into the dipole allowed states with $J^{ex} = 1$ (Γ_5) [157]. Only the latter interact with light forming the exciton–polaritons.

In the material systems investigated, thin epitaxial layers of ZnSe on heterosubstrates and QW structures, these exciton states are further split due to the reduction of crystal symmetry from T_d to D_{2d}. The reason for this symmetry reduction in the case of the epilayers is the strain occurring in the layer due to the lattice mismatch to the substrate [180], while in the QW structures the symmetry is reduced due to the sandwich–like composition from layers of different materials, in our case pure GaAs and mixed crystals of AlAs and GaAs [120]. In the effective mass approximation, the description of the two systems is completely analogous allowing a common group theoretical analysis. Accordingly, the symmetry reduction occurs due to a splitting of the valence band into the heavy hole (hh) and light hole (lh) branches. This induces a splitting of the exciton states, whereby also the exchange interaction has to be taken into account [178].

A simple situation results only in the case when the valence band splitting is large compared to the exchange energy, otherwise exciton states with mixed hh and lh character result. Here we are interested only in the limit of large splitting since in the GaAs system, which is of most interest, the exchange interaction is quite small (0.04 meV, [181]). Then the exciton states split into pure hh states $((n = 1, e - hh), X_{hh})$ with symmetry

$$\Gamma_6 \otimes \Gamma_6 = \Gamma_1 + \Gamma_2 + \Gamma_5 \tag{8.2}$$

and lh states $((n = 1, rme - lh), X_{lh})$

$$\Gamma_6 \otimes \Gamma_7 = \Gamma_3 + \Gamma_4 + \Gamma_5 \ . \tag{8.3}$$

It has to be noted that by the symmetry of the states the coordinate system, to describe e.g. the polarization states, has to be chosen along $x' \| [110]$ and $y' \| [\bar{1}10]$ within the quantum well layer and $z \| [001]$ perpendicular in the growth direction. Then because of the optical selection rules, the dipole transition moments of the hh Γ_5 exciton states are parallel to the x' and y' axes, respectively. The transition moment of the two other states of the hh exciton (Γ_1, Γ_2) is equal to zero and these are therefore optically inactive. In a first approximation, they will be neglected in the following.

From the optical transition moments the coupling matrices of the states with the light field follow according to the procedure given in Sect. 5.2.4, where the case of states with symmetry Γ_5 corresponding to the case of the epilayers and to the QW structures was thoroughly discussed. According to these considerations general relations hold for the contribution of the Rayleigh scattered (RS) part and that of the hot luminescence (HL) with polarization parallel and perpendicular to that of the exciting laser

$$I_\| = I^{RS} + I^{HL} \quad , \quad I_\perp = I^{HL} \ . \tag{8.4}$$

These relations show that the difference in intensities $I_\| - I_\perp$ directly represents the contribution from resonant Rayleigh scattering.

8.1.2 Disorder in Quantum Well Structures

While in the epilayers of ZnSeS mixed crystals the Rayleigh scattering occurs because of the alloy disorder, its occurrence in the GaAs/AlGaAs quantum well structures needs a more profound analysis of the disorder in this system. In an ideal quantum well the electron and hole states and also the excitons form two–dimensional, infinitely extended Bloch states. Therefore the electronic susceptibility is homogeneous in the QW plane. An electromagnetic wave incident on such a system produces according to the considerations in Sect. 5.1 only a reflected wave and no Rayleigh scattering occurs at all.

In contrast, in a real QW, the interfaces between the different materials are never flat on an atomic scale with a sharp compositional transition. Instead, there occur fluctuations in the chemical compostion due to an exchange of the constituent atoms across the interface; these are unavoidable from a thermodynamic point of view [120].

The influence of these compositional fluctuations on the electronic states of the QW is expected to be quite similar to that occurring in bulk semiconductor solid solutions (see [182, 183]). Due to the differences in the atomic energies of the constituent atoms, which are reflected in the different energies of the valence and conduction band in the two materials, the exchange of the atoms at the interface of the heterostructure produces statistically fluctuating attractive and repulsive potentials for the electrons and holes. Due to the finite barrier height, the wavefunctions of electrons and holes extend into the barrier of the QW. As this is in most cases made from a mixed crystal alloy, fluctuations occur, which in addition influence the electronic states.

Quite generally, the scattering of electrons at fluctuating potentials leads to a finite coherence length or mean free path ξ_c of the band states and, with increasing disorder strength, ultimately to the localization of the electronic states [184, 185]. According to the ratio of ξ_c and the exciton Bohr radius a_B^{2d}, which for GaAs quantum wells is of the order of $\simeq 10$ nm, two extreme cases can be distinguished.

1. The coherence length of the band states ξ_c is smaller than the exciton Bohr radius. This case occurs if the disorder scattering is strong, in other words, if the difference in atomic energies of the constituent atoms is large compared to the bandwidth of the electronic states. In most material systems, only a single type of carriers is influenced by the disorder, as e.g. is the case for the hole in II-VI semiconductors with anion substitution, only this type of carrier is localized by the disorder. The other partner which is more or less mobile in the system is then bound by Coulomb attraction thus forming localized exciton states. These have properties similar to excitons bound to impurities and will not be considered here further. Recently, the dynamics of such excitons has been studied using non-linear optical techniques like photon-echoes and four-wave-mixing [186].

2. The coherence length ξ_c is larger than the exciton Bohr radius. This happens if either the disorder–induced scattering is weak or if the effective masses of the carriers are small and hence the bandwidth is large. In this case the excitons are localized as a whole in the fluctuating potentials. This means that a coherent exciton motion occurs only in regions of finite extension determined by the coherence length of the band states. Now in most situtions, the localization energies of these excitons is smaller than their homogeneous linewidth, which is determined by all excitonic relaxation processes. Then the behaviour of these excitons cannot be distinguished from that of free states in a system without disorder as they will be scattered due to the relaxation processes before they reach the borderline of the coherent motion. This type of localized state was therefore designated a *quasi-free* exciton [183].

To describe the dynamics of these quasi-free excitons in a weakly disordered system, the model of a microcrystalline system has been put forward [129]. Here the QW is assumed to be made up from small regions where the exciton states are not disturbed by the disorder, and undergo a Bloch-like coherent motion as is schematically shown in Fig. 8.1. The spatial extensions of these microcrystals is given by the coherence length ξ_c of electron and hole states of the QW. Inside each of these regions, the QW provides in the growth direction an effective localization potential for the electronic states that can be characterized by an effective well width $L_{z,\text{eff}}$ [121]. The differences in the values of $L_{z,\text{eff}}$ across the QW plane result in somewhat different exciton transition energies of the regions leading to the inhomogeneous broadening of the exciton transition. As this mechanism will be effective in even the best QW samples that can be made, this inhomogeneous broadening is always present indicating that localized states govern the optical response of these systems.

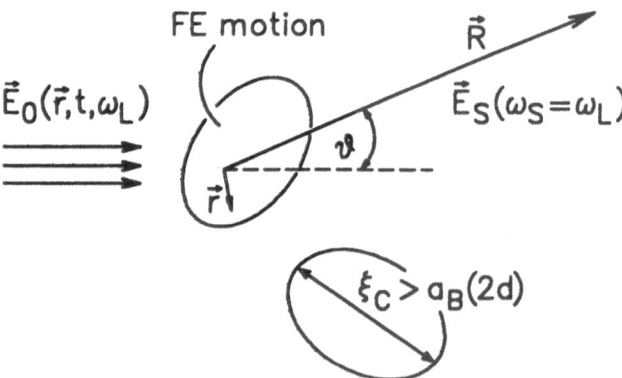

Fig. 8.1 Microcrystalline model of exciton states in quantum well structures showing the regions of coherent exciton motion and the principle of Rayleigh scattering from these states. After [52].

The spatially fluctuating transition energies now result in spatial fluctuations of the resonant electronic susceptibility that leads to the occurrence of resonant Rayleigh scattering in the QW structures. As long as ξ_c is smaller than the wavelength of light the localization of the excitonic states supresses the polariton character also expected for two-dimensional semiconductor systems [187] and the assumptions of the theory of Sect. 5.1 apply. As there is no coherent coupling between the different spatially separated microcrystalline regions, the QW corresponds to a system of many independent atoms [5] and the total scattered intensity is given as the incoherent sum of the scattering from each of these regions.

8.2 Experimental Details

For the measurement of Rayleigh scattering a major problem is posed by the reflected and scattered laser light occurring at the surfaces of the windows that are positioned in the excitation laser beam, as this light is not suppressed by the monochromator. Therefore, neither the direct nor the reflected laser beam must appear in the acceptance cone of the lens collecting the scattered light. A rather elegant and also effective solution is to mount the sample in the cryostat at the Brewster angle to the incident beam and to detect the scattered light at an angle of 90° as schematically shown in Fig. 8.2. Due to the high refractive index of the material (for GaAs $n \simeq 3.5$) the directions of the incident and of the scattered light inside the sample make an angle of only 15° and 4° to the normal of the layer $z' \parallel [001]$, respectively, this geometry

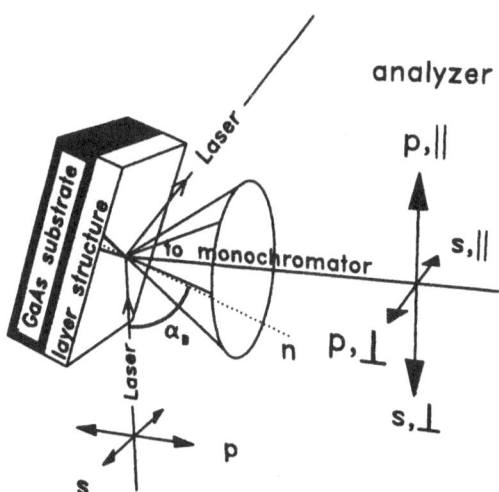

Fig. 8.2 Scattering geometry for the investigation of resonant Rayleigh scattering. From [178].

differs only slightly from the exact 180° backscattering geometry, where these angles would be 0°. The observed scattered light can therefore be assumed to be polarized in the directions parallel to $x'\|[110]$ and $y'\|[\bar{1}10]$ corresponding to the intensities $I_\|$ and $I_\perp$ that have been introduced in the theory. As a further advantage of the excitation under the Brewster angle this geometry allows for incident p-polarization (see Fig. 8.2) according to Fresnel's equations an optimal coupling of the laser light in the layer and a mimimal amount of reflected intensity in the diection of observation. For the Rayleigh scattering measurements, the laser beam is focussed at the sample into a spot as small as possible in order to increase the contribution of the Rayleigh scattered light to the total secondary emission acording to the results of Sect. 5.1. This has the additional advantage that disturbing scattering centres at the surface, e.g. small dust particles, can be avoided.

8.3 Stationary Excitation

8.3.1 Epitaxial Layers of ZnSe/S Mixed Crystals

The simplest way to investigate resonant Rayleigh scattering is to excite the sample with a stationary laser beam with small spectral width that is tuned accross the electronic resonances of the material. However, according to the results of the last section, in this case one can obtain only information about the resonance profile of the Rayleigh process being determined both by the homogeneous and the inhomogeneous linewidth of the states, as the spectral lineshape of the scattering is identical to that of the laser itself. On the other hand, this property of the Rayleigh scattering allows one to discriminate this process from ordinary luminescence, the spectral width of which is determined by the linewidth of the emitting states.

In order to measure the Rayleigh resonance profile, a series of spectra with different excitation photon energies has to be taken, each spectrum scanning the region of the laser line. The result of such a measurement is shown in Fig. 8.3 for an epitaxial layer of mixed $ZnSe_{1-x}S_x$ with a small amount of sulphur ($x \simeq 0.01$) that is grown on a GaAs substrate by metal-organic vapour–phase epitaxy [188].

Comparing the series of Rayleigh spectra with the photoluminescence spectrum of the same sample shown in Fig. 8.4, obviously a dramatic increase in the elastically scattered intensity occurs in the region of the excitonic transitions. This increase depends on the sample and even on the postion of the laser excitation spot on the sample surface and may reach enhancement factors up to 500 under optimum conditions. As expected from the theory (see Sect. 5.1), the spectral width of all Rayleigh scattering lines is the same for each spectrum and identical to the spectral bandwidth of the exciting laser light, which in these measurements was 0.15 meV, while the spectral

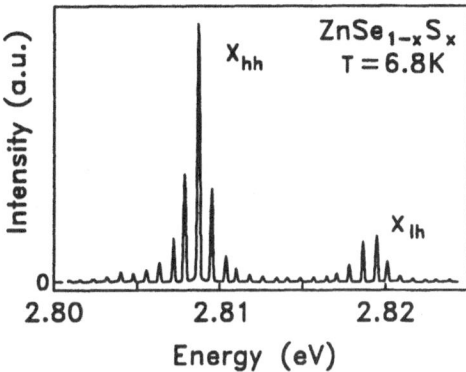

Fig. 8.3
Series of resonance Rayleigh spectra for a $ZnSe_{1-x}S_x$ epitaxial layer on [001] GaAs substrate. Layer thickness 60 nm, T = 6.8 K. From [178].

resolution of the monochromator was less than 0.1 meV. This proves that here we indeed observe a disorder–induced resonance Rayleigh scattering process involving the excitonic states in the epitaxial layer. The information about the temporal dynamics of the exciton states, which is also inherent in the Rayleigh process, of course cannot be obtained from the stationary measurements. Here time-resolved investigations similar to that presented below for GaAs quantum well structures are required; these are presently underway.

Instead of the time-consuming measurement of a series of Rayleigh spectra it is possible to monitor directly the peak intensity of the Rayleigh line. Then one has to change the laser photon energy and the spectral position of the monochromator in identical steps synchronously across the exciton transition. In the experimental setup this was possible because of the high accuracy of the determination of the laser photon energy by the laser monochromator (see

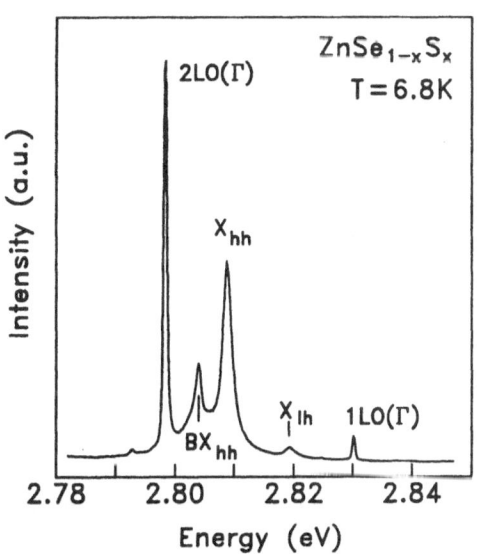

Fig. 8.4
Photoluminescence spectrum of the 60 nm epitaxial layer as in Fig. 8.3. X_{lh}, X_{hh}: Emission lines of the light and heavy hole exciton states. BX_{hh}: Emission line of the donor bound heavy hole exciton. $1LO(\Gamma), 2LO(\Gamma)$: One and two $LO(\Gamma)$ Raman scattering. From [178].

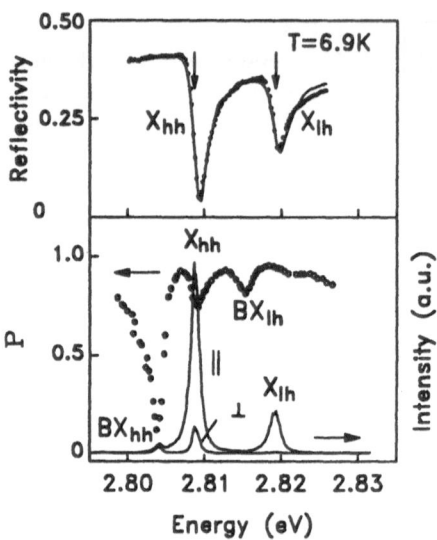

Fig. 8.5
Comparison of the reflection spectrum
(a) with the resonance Rayleigh profile (b) of the 60 nm thick epitaxial $ZnSe_xS_{1-x}$ layer ($x \simeq 0.01$) on GaAs substrate also shown in Figs. 8.3 and 8.4. The arrows in the upper part denote the position of the transverse exciton states as obtained from the analysis of the reflection spectrum. From [178].

Sect. 3.2.3). In this way a continuous curve, the *Rayleigh resonance profile* is obtained.

For the sample already discussed, this Rayleigh resonance profile is shown in Fig. 8.5 together with the conventional reflection spectrum of this sample. Here the Rayleigh intensity was measured with linear polarization both parallel and perpendicular to that of the exciting laser, which itself was p-polarized. Actually, the difference of the two measurements gives the coherent contribution to the Rayleigh process, that is the quantity obtained in the classical treatment, the perpendicular component being the hot luminescence of the exciton states. Here we see that for these extremely thin epilayers the resonances of the Rayleigh scattering occur at the energy positions of the transverse exciton states (arrows). To determine the transverse exciton energies from the reflection spectrum the standard method of Hopfield [111] was used (see [178]). Compared to this tedious procedure, the analysis of the Rayleigh resonance profiles is much simpler. In addition, the measurement of resonance Rayleigh scattering has the advantage that weak electronic resonances like defect bound exciton states not visible in the reflection spectrum can also be studied even in very thin samples. In this context it should be noted that from the results obtained up to now a general rule can be stated concerning the degree of polarization ($P = (I_{\parallel} - I_{\perp})/(I_{\parallel} + I_{\perp})$) for Rayleigh scattering involving free and bound exciton states. For free excitons the polarization degree P is always found to be quite large (compare Fig. 8.5), while for bound states P is small. Whether this effect can be completely explained by the different lifetimes of these exciton states (whereby free states have rather short relaxation times and bound states show a long decay) is at present not clear and needs further investigations.

8.3.2 Stationary Investigations in Quantum Well Structures

Another example for electronic states in a disordered system are excitons in GaAs/AlGaAs QW structures , where actually resonant Rayleigh scattering was investigated for the first time [132, 140].

In order to demonstrate the possibilities of Rayleigh scattering, results will be dicussed that have been obtained in a multiple QW structure consisting of three single wells of pure GaAs with wellwidth $L_z = 9$ nm enclosed between barriers of thickness $L_b = 105$ nm of GaAs/Al$_x$Ga$_{1-x}$As with x = 0.43 [74]. Similar results have been found in other QW samples ranging in L_z from 3 to 15 nm. These samples were grown without growth interruption at the interfaces between GaAs and AlGaAs using molecular beam epitaxy (MBE) by G. Weimann at the FTZ (Darmstadt).

A typical photoluminescence spectrum of the sample with $L_z = 9$ nm at low temperatures is shown in Fig. 8.6. The most prominent peak originates from the recombination of the heavy hole exciton states showing a characteristic doublet structure that indicates a high sample quality. The peak at the low energy side (LE) was previously interpreted as the recombination line of the $(\Gamma_1 + \Gamma_2)$ states [189], whereby the splitting should reflect the electron-hole exchange energy. However, recent results on the excitation power dependence of this line [74] suggest that it is instead due to localized exciton states. From the Rayleigh investigations discussed below the high energy peak at a photon energy of $E_0 = 1.555$ eV can be attributed to quasi-free exciton states (FE) of the disordered system. At still higher photon energies a weak line can be observed that is attributed to the recombination of the light hole exciton states. Due to the larger bandwidth, a distinction between LE and FE states is not possible here.

It is obvious that despite the excellent sample quality, the transitons are inhomogeneously broadened due to disorder at the interface and in the barriers. For the sample shown the inhomogeneous width is $\delta E_{inh} \approx 1.2$ meV

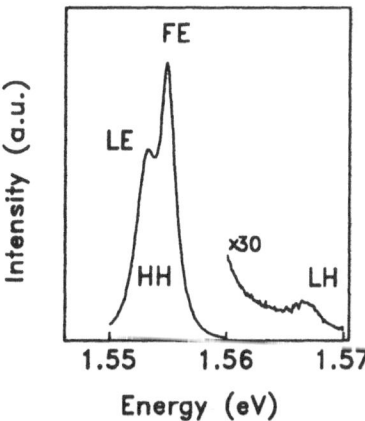

Fig. 8.6
Photoluminescence spectrum of a multiple quantum well sample with $L_z = 9$ nm. LH,HH: light and heavy hole exciton transitions; FE: free excitons, LE: localized states. T = 2 K, excitation power density 900 Wcm^{-2}. From [74]

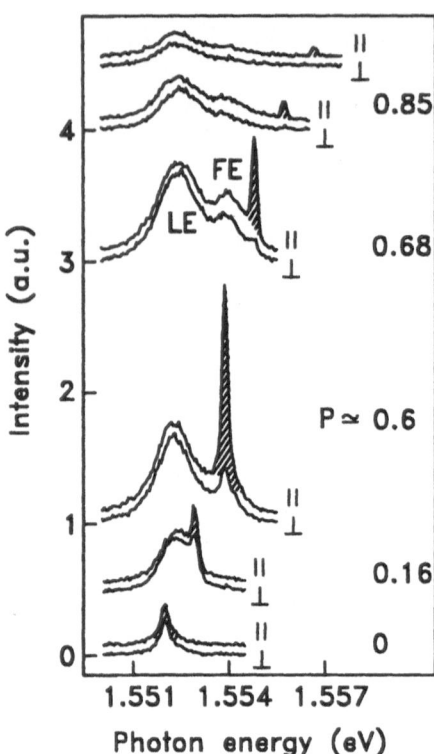

Fig. 8.7
Light scattering spectra from a QW structure resonantly excited in the spectral range of the hh exciton transition. P-polarized excitation, detection polarizations are parallel (∥) and perpendicular (⊥) to that of excititation. $T = 2$ K. P denotes the linear degree of polarization of the scattered light at the peak of the laser line. The hashed area marks the pure Rayleigh contribution. From [140].

(FWHM) as follows from a lineshape analysis, in agreement with the photo-luminescence excitation profile [74]. A Stokes shift between the peak energies in the excitation profile and in the photoluminescence spectrum was not measurable. This usually is taken as an indication of excellent sample quality.

Due to the inhomogeneous broadening of the exciton transitions and the rather weak coupling of the states to the light field in this material we expect to have an ideal model system to study the question of the temporal dependence of disorder–induced Rayleigh scattering. Indeed, it was in similar samples that resonant Rayleigh scattering in a solid was first observed under stationary excititation conditions [132].

A typical example of the resonance behaviour of the Rayleigh process is shown in the series of spectra displayed in Fig. 8.7. The increase of the Rayleigh scattered intensity is clearly observed in tuning across the ($n = 1, e - hh$) exciton transition. The spectral width of the resonance agrees well with the inhomogeneous linewidth of the exciton.

For these measurements the exciting laser beam was focussed to a spot size of about 50 μm diameter at the sample surface. The average power density was reduced well below 0.1 W/cm^2 to avoid saturation effects. The excitation spot was directly observed by a CCD camera system and scanned across the sample surface to a place where the scattered intensity was minimal outside resonance. In this way enhancement factors of about 50 could be achieved.

Compared to the excitation power densities used commonly, e.g. to excite the spectrum shown in Fig. 8.6, the extremely low excitation power changes the appearance of the luminescence spectrum quite drastically. Here the intensity of the LE transition is enhanced compared to the FE line, thus directly proving the saturation of the LE transition and its localized character.

In the measurements shown in Fig. 8.7, the spectral bandwidth of the exciting laser ($\delta E_L < 0.02$ meV) was much smaller than the homogeneous linewidth of the states (see Fig. 8.14). Under these conditions the linewidth of the Rayleigh scattered light is given by that of the laser δE_L as was already discussed for the measurements in ZnSe epilayers. The larger width of 0.25 meV, apparent in the spectra independent of excitation energy, is determined by the spectral resolution of the monochromator, the slits of which had to be opened up for reasons of intensity. Comparing with the data in Fig. 8.12, this is still smaller than the homogeneous linewidth at high excitation energies.

According to the selection rules for the optical transitions of the heavy hole states (see previous section), the spectra shown in Fig. 8.7 have been measured with polarization directions parallel and perpendicular to that of the excitation, which in this case was in the plane of incidence (see Fig. 8.2). As can be seen from the spectra in the Rayleigh forbidden geometry ($I_\perp$) an appreciable amount of depolarized scattering occurs depending on excitation photon energy. This can be expressed quantitatively as a degree of linear polarization ($P = (I_\parallel - I_\perp)/(I_\parallel + I_\perp)$) varying between 0.85 for excitation at the high energy side of the FE line and a complete depolarization in the LE transition. As the resonance enhancement depends on both the homogeneous and the inhomogeneous width of the states, absolute measurements of the scattered intensity would be necessary to derive from P the coherence time of the electronic states. This, however, is not necessary for the time–resolved measurements discussed in the next section.

8.4 Time–Resolved Measurements in GaAs Quantum Wells

8.4.1 Experimental Results

As another example to demonstrate the possibilities of transform–limited time-resolved light scattering (TRLS) compared to conventional absorption and luminescence studies we will now discuss the investigations using time–resolved resonant Rayleigh scattering from excitons in QW structures. Based on the theoretical discussion of the last chapter one expects the resonant contribution to the Rayleigh line to decay with the coherence time of the exciton

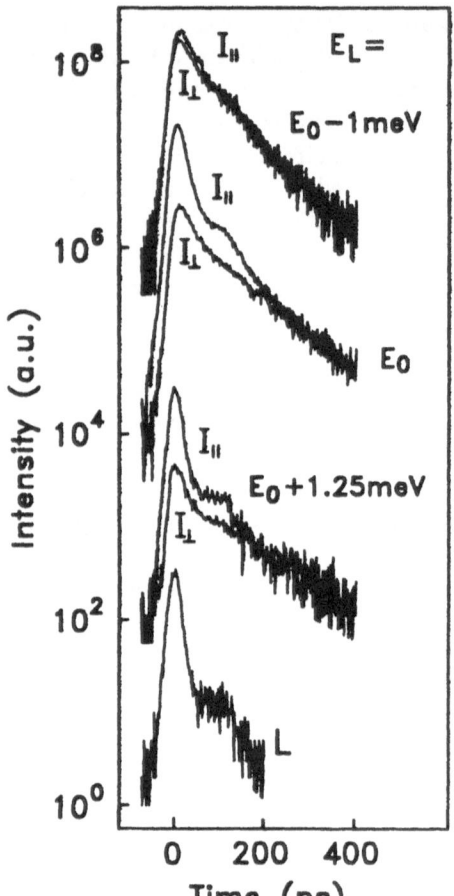

Fig. 8.8
Time dependence of the elastically scattered light after picosecond excitation in a GaAs QW structure with $L_z = 9$ nm. The incident photon energy E_L is tuned across the resonance of the free ($n = 1$, e−hh) exciton state at energy E_0. The scattered light at energy $E_S = E_L$ was detected with linear polarization parallel to x' ($I_\parallel$) and y' ($I_\perp$). L marks the time response of the setup to the laser pulse (duration 5 ps) outside of the resonance. $T = 2$ K.

states. Indeed, it was demonstrated recently for these QW structures that after excitation with picosecond laser pulses and time–resolved detection of the scattered light the Rayleigh process exhibits a finite decay time [140]. In Fig. 8.8 examples are shown of the observed time dependence of the Rayleigh scattered intensity measured in two orthogonal polarization directions for different excitation photon energies tuned across the heavy hole exciton transition. From these data the difference in the temporal behaviour between the two polarizations is clearly visible. The parallel polarized component, which according to theory contains the proper Rayleigh process, is strongly enhanced during the first 50 ps compared to the perpendicular intensity comprising only the phase relaxed contributions to the scattered light. The difference between the two components has to be identified with the Rayleigh signal. Comparing the temporal dependence of the scattered light with the response of the setup to the laser pulse (lower curve in Fig. 8.8) reveals the delayed decay of the resonantly enhanced Rayleigh scattering. According to the autocorrelation signal, the laser pulse used in these measurements had a pulse width of about 5 ps and a spectral bandwidth of 0.3 meV. The slow decay (time constant ~ 150 ps) that is found both in $I_{\parallel}$ and in $I_{\perp}$ can be understood as a radiative recombination from the localized LE states that are occupied by exciton capture [74] or from FE states that have completely relaxed to the minima of the fluctuating potentials. Most important is the variation of this decay time with excitation photon energy implying that no thermal equilibrium exists within the duration of the emission process.

In order to display more clearly the temporal dependence of the Rayleigh process, the intensity differences $I_{\parallel} - I_{\perp}$ are plotted on an enlarged time scale for different excitation photon energies in Fig. 8.9. Here each curve is normalized to its maximum and compared directly with the system response to the laser pulse outside the resonance (dotted line). Comparing the time dependences, it is obvious that indeed a small time delay exists between the response to the laser pulse and the Rayleigh scattered light. This delay increases with increasing decay time as expected from theory (see Sect. 5.3.5). In these experiments, care has to be taken in comparing the laser response with the scattered light, as most often the laser pulse is attenuated using neutral density filters made from absorbing glass. Here we have found that these glass filters cause a time shift of the order of 10 ps due to pulse propagation effects that depends on the type of material and the filter thickness. Therefore in the experiments shown in Fig. 8.9 all measurements were performed with the same neutral density filter inserted in the excitation beam allowing a direct comparison of the temporal dependence. In this way the delay of the Rayleigh process is directly visible. Above the resonance at the peak of the excitation and luminescence spectrum with energy E_0, the decay of the Rayleigh process is fast, while it slows down with decreasing excitation energy. This directly reflects the energy dependence of the excitonic relaxation rates.

To obtain the relevant relaxation times from these measurements, the scattered intensities were analyzed in the framework of the transform–limited theory. In order to do this, first the resonantly excited exciton states have to be characterized. For an ideal QW the incident light wave couples with all exciton states for which the component of wavevector in the QW plane $K_{\parallel}$ is equal to that of the light wavevector in the plane (see Fig. 8.11). This approximately still is valid in the microcrystalline model for the FE states in a disordered system. Thus, for perpendicular incidence, which can be assumed to be realized here, only exciton states with $K_{\parallel} = 0$ interact with the exciting laser light and give rise to the scattered field as depicted schematically in Fig. 8.10. Therefore, the sum in Eqs. (5.121) and (5.131) has to be performed only over the two degenerate Γ_5 states $|\alpha\rangle$ with $\alpha = x', y'$. This result then has to be summed over the inhomogeneous distribution of exciton energies. To make the fit quantitative, additional slowly decaying contributions have to

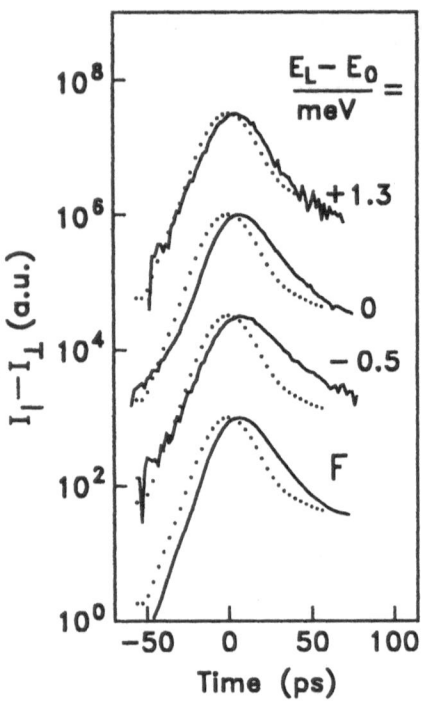

Fig. 8.9
Time dependence of the polarized part of Rayleigh scattering for different excitation photon energies E_L referred to the maximum of the FE emission at $E_0 = 1.555$ eV. The dotted curves represent the temporal response to the laser pulse outside resonance, the curve marked by F is the result of the calculation by the transform-limited theory for the measured curve at $E_L = E_0$.

be included that take into account the relaxation into the LE or localized FE states with rate Γ_{capt} (in Fig. 8.10 denoted both as LE) and their radiative decay with lifetime τ_{LE}. These components have been modelled by a double exponential decay function with rise time $1/\Gamma_{capt}$ and decay time τ_{LE}. For all states we assume the same homogeneous linewidth Γ in the calculations. The spectral width of the Fabry Perot resonator used as detection filter and the time constant of the *single–sided–exponential* shaped laser pulse were taken to be $\gamma_L = 2.5 \cdot 10^{11}$ s^{-1} and $\gamma_S = 3 \cdot 10^{11}$ s^{-1} in agreement with the experimental conditions.

The result of a typical fit of the polarized intensities under these assumptions is shown in Fig. 8.11. Obviously, the measured time response can be

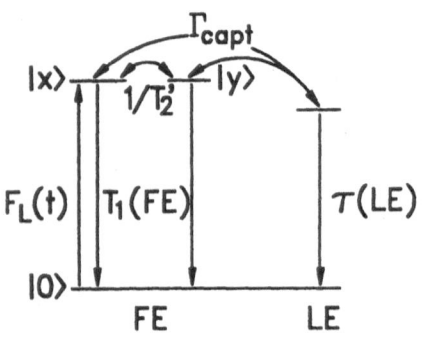

Fig. 8.10
Schematic model of the quantum well states used to analyze the Rayleigh scattering.

described very well by the theory. In particular, the shift between $I_\parallel$ and $I_\perp$ at early times, which cannot be described by the usual rate equation theory, is reproduced correctly. Only during the first 20 ps do some small deviations remain for the parallel component; these can be traced back, however, to the difference in the temporal behaviour of a Fabry Perot spectral filter from that of the grating monochromator actually used in the experiment (see Sect. 2.1.3). The true Rayleigh component given by $I_\parallel - I_\perp$ depends only on the coherence time of the exciton states which is found to be of the order of ten picoseconds. Due to this rather small value, the fitting procedure is quite critical. In particular, the residual time broadening of the photodetector has to be modelled accurately allowing the transform-limited characteristics of the setup to be taken properly into account. The achievable quality of the fit is demonstrated by the lower curves in Fig. 8.9 (marked F) which agree almost quantitatively with the experimental results (curve at $E_L = E_0$).

The fitting procedure of all decay curves allows one to deduce both the coherence time or homogeneous linewidth of the exciton states $\tau_{\rm coh} = 2/\Gamma$ and the energy relaxation time $T_1 = 1/\Gamma_{rmER}$. Here it has to be pointed out that according to the theoretical discussion in Chap. 5, the homogeneous linewidth actually corresponds to the upper state coherence time, in particular to the spin relaxation time $1/\gamma_2$ because of excitation with linearly polarized light (see Sect. 5.2.4). Only for simple multi-level systems, like those discussed in Sect. 5.3.5 and used here to analyze the experiments, is this time identical to

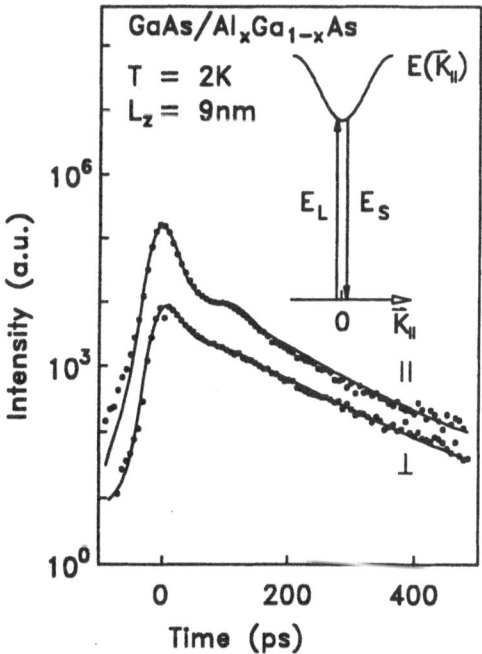

Fig. 8.11
Time dependence of Rayleigh scattering: Comparison of the results of the transform–limited theory with the experiments. All experimental conditions are the same as in Fig. 8.8. The inset shows schematically the two–dimensional exciton dispersion $E(K_\parallel)$ and the resonant Rayleigh process.

the decay time measured in four-wave-mixing experiments, in other cases the two times may be quite different.

For the curves displayed in Fig. 8.11 these times are $\tau_{\text{coh}} = 18$ ps and $T_1 = 12$ ps. From these data the pure dephasing time $T_2' = 2/w = 70$ ps can be deduced. The decay time of the LE states results as $\tau_{\text{LE}} = 120$ ps.

Recently, several other attempts to investigate the relaxation dynamics of excitons in GaAs quantum well structures by time-resolved fluorescence spectroscopy have been published [53, 190]. While the experimental results are almost identical to those discussed here, all these authors apply the usual rate equation approach to analyze their experiments claiming not to see any true Rayleigh scattering effects, although the experimental conditions clearly correspond to a transform-limited situation.

As was pointed out in the discussion of the transform-limited theory of light scattering, the rate equation approach is equivalent to the white-light limit of excitation and detection. In this limit, however, resonance fluorescence (understood as absorption followed by emission) and scattering as a one step process cannot be distinguished from each other and describe the same physical phenomenon. In contrast, invoking *spectral resolution*, as is done under transform-limited conditions, coherent scattering and the incoherent hot luminescence can be discriminated as different processes. In particular, the spectral width of the Rayleigh scattering depends on that of the exciting laser, as exemplified in Fig. 8.7 and also observed by other groups [190].

8.4.2 Energy and Temperature Dependence of Relaxation Times

As was put forward in the last section, the analysis of the time dependences of the polarized resonance Rayleigh signal allows one to deduce the phase and energy relaxation times of the exciton states. The results obtained for a series of excitation energies in resonance with the hh transition are displayed in Fig. 8.12. Tuning across the exciton transition, both τ_{coh} and T_1 change in the same way from rather long values at the low energy side of the FE line ($\tau_{\text{coh}} = 20$ ps) to very short times (below 5 ps for τ_{coh}) at the high energy side. Due to the 20 ps time resolution of the photomultipler, this time range represents the lower limit of a meaningful determination of the decay times presently possible by TRLS.

The values found for τ_{coh} agree with those found earlier by the technique of degenerate, time–resolved four–wave–mixing on similar samples [27, 30]. In particular, this agreement in the coherence times shows that the coherence between exciton states in spatially separated microcrystalline regions is indeed negligible, one of the key assumptions of the treatment of Rayleigh scattering in QW structures. Otherwise, the phase relaxation would be governed by the full inhomogeneous linewidth of the transitions which amounts to about 1.25 meV corresponding to a decay time of less than one picosecond for the scattered signal.

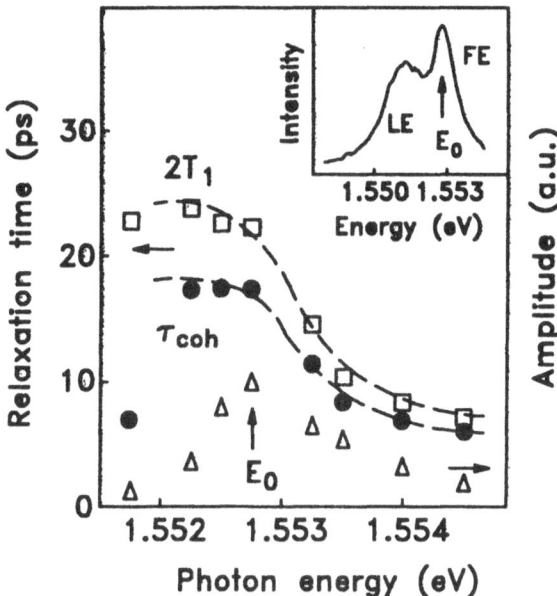

Fig. 8.12 Dependence of the energy and phase relaxation times on excitation photon energy derived from the analysis of time-dependent Rayleigh scattering. The full circles give the phase coherence times τ_{coh}, while the open squares give twice the energy relaxation time T_1 and the open triangles give the scattering amplitudes. The dashed line is a guide for the eye only. The inset shows the photoluminescence spectrum of the sample. FE : free exciton transition with maximum at E_0. LE : localized states. From [74].

As is obvious from the small difference between τ_{coh} and $2T_1$ in the energy region of the FE transition, the pure dephasing relaxation with rate w gives here only a small contribution to the homogeneous linewidth of about $w = 0.03 - 0.04 \cdot 10^{12} \text{s}^{-1}$ which is nearly energy independent. The phase relaxation of the FE states, therefore, is determined by energy relaxation processes that depopulate the exciton states. The quantitative results for Γ^{ER} and the reciprocal of the decay time τ_{LE} of the LE states are summarized in Fig. 8.14 to allow a more detailed discussion of the origin of these relaxation processes.

In contrast, the coherence time of the LE states is short and could not be resolved with our time–resolution ($\tau_{coh} < 5\,\text{ps}$). Therefore, because of the long energy relaxation time of about 100 ps mostly pure phase relaxation processes contribute to the homogeneous linewidth of the LE states. A similar fast phase relaxation of localized exciton states was recently also observed for localized states in GaAs/AlAs mixed crystals [186]. This suggests that any explanation of this fast dephasing has to take the peculiarities of the electronic states in this system into account. In the GaAs/AlAs system discontinuities occur both in the valence and in the conduction band. Therefore,

both electrons and holes are scattered by the disorder–induced potential fluctuations and can be localized. This is in contrast to solid solutions of II–VI
compounds with anion substitution, where the fluctuations in chemical composition affect only the valence bands localizing the holes in regions of the
order of the lattice constant [182], while the electron states remain essentially
free. Here rather long phase relaxation times of several hundreds of picoseconds have been observed [186]. The LE states of the QW structures most
probably also originate from a localized hole binding an electron by Coulomb
interaction [74]. Due to the rather weak localization potential the spatial extent of these hole states is much larger than a lattice constant. In the bound
exciton state both particles are then scattered by the disorder potential. This
is just the situation where according to recent theoretical investigations fast
phase relaxation occurs [191].

The amplitude of the resonant Rayleigh process indicated by the open
triangles in Fig. 8.12 follows closely the FE emission lineshape (inset) reaching a maximum at the peak of the exciton line E_0, as expected from theory.
However, Rayleigh scattering occurs also at the high energy side of the FE
line. This indicates that the lineshape of exciton absorption, to which the
amplitude of the Rayleigh process has to be proportional, is highly asymmet-
ric. This effect can be understood in the microcrystalline model by noticing
that due to the finite extent of the wavefunction the k-vector of the exciton
states is not well defined and optical absorption is also possible for states
at non–zero wavevector at energies $\epsilon > 0$. Recently, such a behaviour of
the optical absorption of excitons under the influence of a disorder potential
was calculated in a one–dimensional model and could be traced back to the
translational motion of the excitons averaging over the potential fluctuations
[117]. To clarify this interesting relation further measurements, in particular
in samples showing different amount of disorder, are planned.

In order to explore the origin of the excitonic depopulation processes,
investigations of the temperature dependence of the homogeneous linewidth
have been performed. Such investigations have been done previously by nonlinear four-wave mixing studies of the exciton dynamics [27]. From these experiments it was deduced that the linewidth shows a linear increase with
temperature, whereby the slope depends on the QW width L_z, amounting for
$L_z \simeq 15$ nm to about 5 μeV/K.

Independently, from time-resolved photoluminescence it was found that
the decay time of the excitonic emission excited above the band gap increases
with temperature [192]. This was related to the specific ability of excitonic
states to form a coherent superposition state extending over many elementary
cells of the crystal. The radiative lifetime of such a coherent exciton state
should be proportional to the number of coherently excited elementary cells,
an effect long known for localized states as the 'giant oscillator strength'
[193]. The region of this coherent exciton motion is limited by the relaxation
processes leading to a dephasing of the excitonic wavefunction and can be

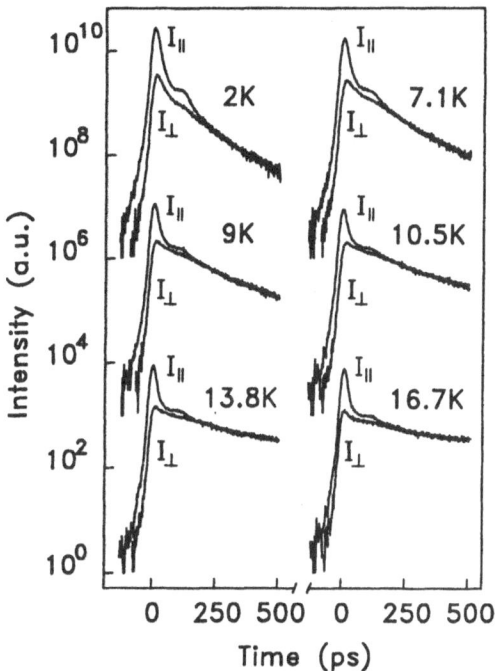

Fig. 8.13
Time dependence of elastically scattered light after pulsed excitation in resonance with the hh-exciton state in QWs for various temperatures. Shown are the intensities with polarization parallel ($I_\|$) and perpendicular ($I_\perp$) to the p-polarized laser pulse with $E_L = E_0$. From [74].

characterized by an excitonic coherence length ξ_{ex}. Then the increase in the excitonic decay time, if interpreted as a radiative decay, can be understood qualitatively as due to the decrease of the coherence length correlated with the increase in homogeneous linewidth [192]. Recently, a similar dependence of the decay time of excitons in QW structures was found for the exciton density which also influences the excitonic dephasing by exciton-exciton collisions [194]. These observations seem to substantiate this picture of exciton states. All these studies, however, have been performed by different experimental techniques making a direct comparison somewhat problematic.

The method of TLRS now offers the unique possibility to measure the phase relaxation processes, i.e. the homogeneous linewidth and the decay time of the relaxed exciton states in the same setup, thus avoiding all the ambiguities connected with a change in experimental methods. A series of measurements of the time dependence of resonance Rayleigh scattering for different temperatures is displayed in Fig. 8.13. The curves clearly show the expected behaviour in which, with increasing temperature, the phase relaxation times get shorter, while the decay at longer times slows down. In our interpretation, however, this slow decay originates from the LE states and not from the free excitons.

The results of the analysis of these measurements is depicted in Fig. 8.15, where the total relaxation rate $\Gamma = 2/\tau_{coh}$ is plotted together with the reciprocal of the LE decay time τ_{LE}.

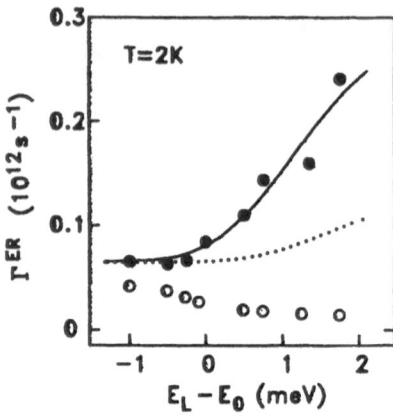

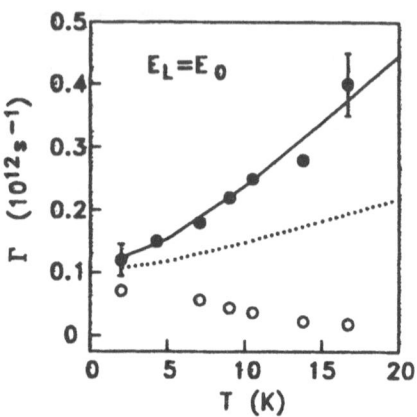

Fig. 8.14 Energy relaxation rates $\Gamma^{ER} = 1/T_1$ (full circles) and reciprocal of LE lifetime τ_{LE} (open circles) plotted versus excitation photon energy E_L as obtained from the analysis of the Rayleigh scattering. The lines are the results of a theoretical calculation using $M_{hh} = 1.03m_0$ (full line) and $M_{hh} = 0.234m_0$ (dotted line) as discussed in the text. E_0 denotes the energy of the maximum of the photoluminescence intensity.

Fig. 8.15 Temperature dependence of the total phase relaxation rate Γ of the FE states (full circles) and the reciprocal of the LE lifetime τ_{coh} (open circles) as obtained from the measurements shown in Fig. 8.13. The full and dotted lines are the results of a theoretical calculation of the relaxation processes using the same exciton masses as in Fig. 8.14. For details, see text.

As processes responsible for the energy and temperature dependence of the relaxation rate four different types have to be taken into account.

Scattering by Acoustic Phonons. According to the discussion in Sect. 4.4, deformation potential scattering by LA phonons is the dominant process. For a quantitative comparison with the experiments, the inhomogeneous broadening of the exciton transition has to be taken into account. For a sharp excitation photon energy E_L, several exciton states in different microregions are excited differing in their kinetic energy. Therefore, the experimentally obtained relaxation rates correspond to an average over the inhomogenous distribution. Using an inhomogeneous broadening of 1.25 meV FWHM, from the results for Γ_{ac} shown in Fig. 4.8, the total phonon scattering rate comprising all relaxation mechanisms is obtained (dotted line in Fig. 8.14). Here for the heavy hole exciton an effective mass of $M_{hh} = 0.234m_0$ was used corresponding to $m_e = 0.069m_0$ and $m_{hh} = 0.165m_0$. While the general shape of both the energy and temperature dependence seems to agree with the experimental results, quantitatively, the rates obtained are about a factor of five too small. Recently, by fitting exciton diffusion experiments in somewhat thicker samples to a semi–empirical formula [195, 196] similar large relaxation rates were found.

As all other quantities are fairly well known, this indicates that the actual value for the exciton translational mass might be quite different from the sum of electron and hole masses. A similar effect is well known from the bulk III-V semiconductors, where due to the Coulomb interaction the exciton masses are always to be found much larger than the sum of the band masses [123]. Assuming in a first approximation that the QW exciton states result from the bulk states by a suitably chosen symmetry-breaking external potential similar to the effect of an external strain, it can be shown that the heavy and light hole exciton dispersions are merely shifted, the translational masses changing only slightly. Taking the exciton masses determined by resonant Brillouin scattering experiments [113], an average value for the in-plane heavy hole exciton mass $M_{hh} = 1.03m_0$ is obtained. Calculation of the acoustic phonon scattering rates with this exciton mass gives the full lines in Fig. 8.14, which agree quantitavely with the experimental results, if one adds an energy-independent depopulation rate of $\Gamma_0^{ER} = 0.06 \cdot 10^{12}s^{-1}$. This rate then has to be due to the other relaxation channels discussed in detail below.

As shown in Fig. 8.15, the temperature dependence of Γ can also be explained using this effective mass value. Here a temperature–independent dephasing rate $\Gamma_0 = 0.095 \cdot 10^{12}s^{-1}$ also has to be taken into account. This rate comprises both Γ_0^{ER} and an almost temperature–independent pure dephasing rate $w = 0.035 \cdot 10^{12}s^{-1}$.

Capture of Free Excitons to Localized States. From the photoluminescence spectrum it can be deduced that mostly LE states are involved in the capture process. In our analysis of the exciton dynamics (see Fig. 8.10) this relaxation channel is already included by the term with Γ^{capt}. From the fitting procedure, capture rates of $0.01 \cdot 10^{12}s^{-1}$ emerge, which are too small to explain Γ_0^{ER} [74].

Relaxation of the Optically Allowed Γ_5 Exciton States to the Optically Forbidden $\Gamma_1 + \Gamma_2$ States by Hole Spin Flipping. This type of process was discussed extensively in [197] and leads to a depopulation of the optically active FE states in the same way as capture processes. Due to the fast hole spin relaxation usually found in bulk III-V semiconductors, this process is expected to be quite fast, but can be strongly suppressed by the exchange interaction between electron and hole in the exciton states. In a recent time-resolved luminescence study in samples similar to the ones used here [197], a spin-flip rate of about $0.01 \cdot 10^{12}s^{-1}$ was deduced, which also is too small to explain the observed relaxation rate. Different from the LE state, the emission from the triplet states is forbidden, so that they show up only after thermal activation as additional free exciton emission making the analysis somewhat uncertain. A more reliable measurement of the spin-flip scattering would require direct observation of the triplet emission. This can be achieved by the application of a magnetic field mixing singlet and triplet states. Preliminary experiments performed by us [198], however, have shown that for magnetic fields up to 4 T, no change in the relaxation dynamics is observable.

Radiative Decay of the Exciton States. This process represents the ultimate relaxation channel for excitons and was discussed quite a lot in the recent literature. While the model of giant oscillator strength, explained above, in essence requires a localized character of the exciton states, in an ideal QW the excitons actually form two-dimensional (2d) polariton states [199] due to the strong coupling with the external photon field. Thereby, for a proper description, the whole system has to be treated as a dielectric wave–guide structure [187]. In contrast to three-dimensional polaritons, 2d-polaritons exist for in-plane wavectors $|K_{\parallel}| < k_0$ (with $k_0 = n\omega/c$ the photon wavevector inside the QW and n the index of refraction) in the form of radiative modes leading to an ultrafast radiative decay of the order of several tens of picoseconds [187, 199] for these states. The radiatative transition rate of the polariton states at $K_{\parallel} = 0$ is thereby given by [199]

$$\Gamma_{\mathrm{rad}} = \frac{\pi e^2}{4\pi\varepsilon_0 nm_0c}\tilde{f}_{xy} \ . \tag{8.5}$$

Inserting the values for $n = 3.5$ and $\tilde{f}_{xy} = 5\cdot 10^{-3}$ nm^{-2} for the oscillator strength appropriate for the hh exciton in a $L_z = 9$ nm QW structure, $\Gamma_{\mathrm{rad}} = 0.037\cdot 10^{12}$s^{-1} is obtained, which is of the right order of magnitude to explain the observed relaxation rates.

In real samples, however, due to the always present disorder, the fully coherent 2d-polariton states undergo strong elastic scattering which mixes states with different in-plane wavevector resulting in the quasi-free character of the excitonic states in such a structure.

To obtain the radiative decay of these quasi-free exciton states, the polariton decay rates have to be suitably averaged. Assuming an exponential localization of the exciton translational motion, the coherence length ξ_{ex} is given by half the decay length of the excitonic wavefunction down to $1/e$ of its initial amplitude. The decay length can be expressed as an imaginary part of the excitonic wavevector $\Im m(\tilde{k}) = 1/2\xi_{\mathrm{ex}}$. This imaginary part of $\tilde{k}$ can be obtained from the complex excitonic self-energy $\tilde{\epsilon} = \epsilon - i\hbar/2\Gamma$ by assuming an isotropic and parabolic dispersion relation

$$\epsilon(k) = \frac{\hbar^2 k^2}{2M^*} \tag{8.6}$$

to be

$$\Im m(\tilde{k}) = \left[\frac{\sqrt{2M^*}}{\hbar c}\sin\left(\arctan\left(\frac{\hbar\Gamma}{2\epsilon}\right)/2\right)\left(\epsilon^2 + \hbar\Gamma^2/4\right)^{1/4}\right] \tag{8.7}$$

with exciton mass M^* and total phase relaxation rate Γ. This relation shows that the coherence length depends on both the homogeneous linewidth and the exciton kinetic energy. In the limiting case of $\epsilon = 0$ this relation is identical to that postulated previously [192], while for large values of ϵ a relation equivalent to

$$\tau_{\text{coh}} = \xi_{\text{coh}}/v_g \tag{8.8}$$

results. This latter relation can be derived using the following simple arguments [74]: A wavepacket of excitons with energy ϵ travels with a group velocity $v_g = d\omega/dk$ given by

$$v_g = \sqrt{2\epsilon/M^*} \ . \tag{8.9}$$

Then a well–defined phase relation between the exciton waves is destroyed within the coherence time τ_{coh} or, in other words, within the coherence length ξ_{coh}. Relating these quantities just leads to Eq. (8.8). Using a value for the homogeneous linewidth of about 0.1 meV appropriate at low temperature near the exciton energy E_0 (see Fig. 8.15), and an exciton mass of $M_{hh} = 1.03m_0$ an excitonic coherence length of about $\xi_{\text{ex}} = 40$ nm is obtained. Being much larger than the exciton Bohr radius, this length has the right order of magnitude to justify a description of the excitons with the microcrystalline model. The averaging of Γ_{rad} over the polariton states with different wavevector will reduce the radiative transition rate of the quasi-free exciton states compared to the coherent polariton value, the result depending on the product of the light vector $k_0 \simeq 1/36$ nm^{-1} with the coherence length ξ_{ex}. Here it can be shown that for $k_0 \cdot \xi_{\text{ex}} > 0.5$ the reduction is negligible resulting in a radiative decay time of about 50 ps. Only for much smaller localization length of the order of the Bohr radius does the decay time of the exciton states increase up to several hundred picoseconds. Obviously, to explain the decay times of the LE states such small coherence lengths are needed in agreement with their localized nature [200]. The assumption of thermal equilibrium for these states that is implicit in the discussion of the long decay time in the previous work [192, 53] seems not to be justified as the experiments clearly show that the decay time depends on the emission photon energy. This would be impossible under thermal equilibrium. On the other hand, this implies that the energy and temperature dependence of the decay time of the LE states has to originate in the dephasing processes of the LE states themselves, but a profound understanding would require more detailed measurements.

In conclusion, we have shown that time-resolved light scattering can be used to derive reliable values for the rates of the various excitonic relaxation processes. From the energy and temperature dependence it was posssible to explain the nature of the processes. At low energies, radiative decay and exciton trapping determine the lifetime, the radiative decay time being of the order of 30 ps, comparable to that expected for almost free exciton-polariton states. With increasing exciton kinetic energy, scattering by acoustic phonons becomes more and more important, whereby deformation potential scattering by LA phonons can quantitatively explain both the observed energy and temperature dependence. The pure dephasing relaxation rate is found to be almost energy and temperature independent suggesting as its origin scattering induced by the static disorder found in these QW structures due to interface roughness and concentration fluctuations in the barriers.

9. Outlook

9.1 Further Investigations

While in the foregoing chapters resonant light scattering from representative exciton systems was discussed with emphasis on the fundamental principles, in this section some further investigations will be briefly presented. These are quantum beat experiments with localized excitons bound to impurities in a II–VI compound, CdS, and first results of time–resolved investigations of resonant Brillouin scattering in ZnSe epitaxial layers. These open up new possibilities for studying the dynamics of exciton–polaritons in semiconductors.

9.1.1 Quantum Beats Involving the I_3 Exciton in CdS

As another example of quantum–beat spectroscopy of excitons, experiments involving an exciton bound to an ionized donor (D^+, X complex) in CdS, the transition commonly named I_3 [201], will be presented, where the high precision in determining the energy splittings between the resonant levels allowed refined magneto–optical parameters to be deduced [90].

As depicted in the inset of Fig. 9.1, the electronic structure of the (D^+, X) states closely resembles that of the free $\Gamma_9 \times \Gamma_7$ exciton with a Γ_1 ground state and a fourfold excited state with symmetries Γ_5 (optically dipole–allowed states Φ_1, Φ_2) and Γ_6 (dipole–forbidden states Φ_3, Φ_4) [201]. The behaviour of the transition energies and optical matrix elements in a magnetic field is governed by a Zeeman Hamiltonian, which is similar to that given by Eq. (6.36) with $g_v^\perp = 0$ due to symmetry [201].

The CdS sample was of cylindrical shape with the crystallographic c–axis along the cylinder axis. The direction of the magnetic field and the c–axis enclose an angle Θ, variable by rotating the sample, thus defining a plane of reference. The experiments were performed in Faraday geometry in a nearly backscattering configuration, whereby the resonance fluorescence was excited through the crystal surface with its normal parallel to the c–axis. Figure 9.1 shows examples of the observed time dependence of the I_3 intensities at zero magnetic field and at $B = 2$ T. Here excitation with linearly polarized light

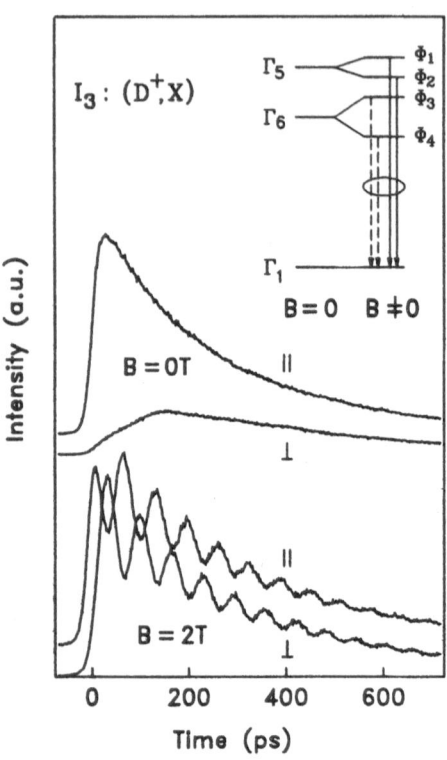

Fig. 9.1

Time dependence of polarized 1LO resonance fluorescence line of an exciton bound to an ionized donor (D^+, X) (I_3) at zero magnetic field and at $B = 2\,\mathrm{T}$ for $\Theta = 0°$. Excitation is with light linearly polarized at 45° to the plane of reference (see text). The scattered light is linearly polarized at 45° ($\parallel$) and $-45°$ ($\perp$). Zero of time scale is given by the system response to the laser pulse (not shown). The inset shows the energy level scheme and magnetic field splitting of the (D^+, X) states with the optical transitions represented by arrows, the dashed ones being only allowed for a magnetic field inclined to the c–axis ($\Theta \neq 0°$). The transitions for which quantum beats occur are grouped together by circles.

was accomplished in resonance with the exciton transition, while the scattered light was detected in the corresponding 1 LO–replica of the line to avoid any disturbing effect of elastically scattered light at the exciting laser frequency.

For the occurrence of quantum beats from the split excited states, the optical transitions connecting at least two of the excited states with the same ground state level have to be allowed. If, due to appropriate choice of light polarizations, only a single state is excited, we expect to observe no beating. Instead, according to the theory of light scattering in Chap. 5, a depolarization of the scattered light occurs. Working out the selection rules, it follows that for the I_3 line, at $\Theta = 0°$, beats occur only between the Γ_5 sublevels, while for $\Theta \neq 0°$ all four transitions may interfere. For the (D^+, X) exciton states we therefore expect at zero field in the scattered light a fast Raman–like component for detection with polarization parallel to that of the exciting laser and a slow luminescence–like component in the orthogonal polarization state, irrespective of excitation polarization. In fact, however, we found for most polarization directions temporal oscillations of the scattered intensity in the two orthogonal detection polarizations, indicating that even at zero field the (D^+, X) exciton states are split by residual strain in the sample. As implied by the absence of any beating for an excitation polarization under 45° to the plane of reference (decay curves for $B = 0$ T displayed in Fig. 9.1) only one of the excitonic eigenstates is excited for this case. Since dephasing effects are small, by applying a magnetic field $B\|c$ beat patterns are observable, extending over the whole lifetime of the states. In the chosen geometry, the beating involves only transitions to the states Φ_1 and Φ_2 and, as shown in Fig. 9.1 (bottom), occurs in both parallel and perpendicular polarization, being phase shifted by 180° as is expected from theory. Both the lifetime of 220 ps and the coherence time of about 300 ps, which were obtained from the damping of the beats and found to be independent of magnetic field, are consistent with the depolarization behaviour at zero magnetic field.

For the chosen excitation and detection geometry, the only parameter (besides the strain splitting) which determines the magnetic field dependence of the splitting of the exciton states is the difference between electron and hole g–factors. From analyzing a whole series of quantum beat signals for varying magnetic fields a value of $|g_c - g_v^\|| = 0.5$ was obtained. This shows that for the I_3 state the electron and hole g–factors are quite different, a result which could not be obtained by conventional magneto–optics [201], because of the large inhomogeneous broadening masking the splitting of the states. As the difference in g–factors is almost equal to that found for the free A–exciton state [202], we conclude that the binding of the exciton to the ionized donor causes only a small change in the electronic wavefunctions of the states.

9.1.2 Time–Resolved Resonant Brillouin Scattering in ZnSe

In Chap. 7, Raman scattering from exciton–polaritons involving optical phonons was studied in the material Cu_2O. Another important scattering process of exciton–polaritons is provided by the interaction with long–wavelength acoustic phonons. Ever since its theoretical prediction by *Brenig, Zeyher* and *Birman* [203] and the experimental verification by *Ulbrich* and *Weisbuch* [204], *resonant Brillouin scattering* (RBS) has been a very powerful method to investigate the dispersion relations of exciton–polaritons by employing the dispersive character of the acoustic phonons (for a review see e.g. [113]). However, both the measurement and the theoretical prediction of the intensities of RBS lines has remained a formidable task and at the present stage a rather unsatisfactory state of affairs. This is due to the fact that the strength of the exciton–phonon interactions, the details of the exciton–polariton wavefunctions, the boundary condition at the crystal surface (the famous ABC problem) and the temporal dynamics, i.e., the damping of the states, all enter in the problem (see e.g. [113]).

Here too, the method of time–resolved light scattering will lead to more easily interpretable experimental results, as it allows one to separate the strength of the scattering from the temporal dynamics. The first study of time–resolved RBS was recently performed by us using high quality layers of ZnSe grown by MOVPE on a GaAs substrate (for details see Sect. 8.3.1) [48].

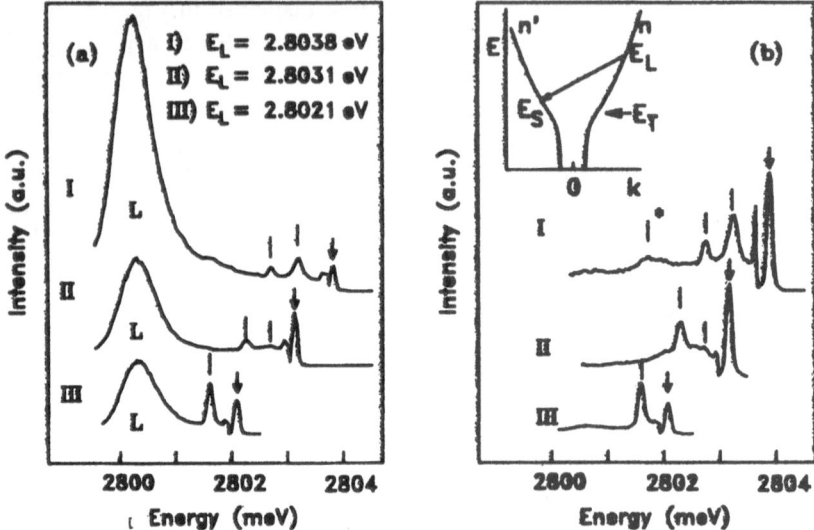

Fig. 9.2 (a): Luminescence and Brillouin scattering spectra under resonant excitation in the 1s exciton transition of an ZnSe epitaxial layer. The excitation photon energies E_L as given are marked by arrows, the Brillouin lines at energies E_S by vertical lines. L denotes the exciton–polariton luminescence. $T = 2$ K. (b): Spectra as in (a), but taken with time sampling of the scattered light (width of time window 150 ps around the system response, see Fig.9.3). The line marked * is a two–phonon scattering process. Inset: Schematic dispersion of polariton branch n and n' showing a Brillouin backscattering process.

The optical measurements at $T = 2$ K were performed in a 90° scattering geometry as shown in Fig. 8.2. This corresponds to nearly 180° backscattering with the propagation directions of incident and scattered light inclined at angles of 20° and 7° to the surface normal. The excitation laser pulse duration was enlarged to about 33 ps by inserting an etalon in the dye laser resonator giving a spectral width of about 40 μeV (see Fig. 3.6) sufficient to spectrally resolve the RBS lines under transform–limited conditions.

For above–bandgap excitation the emission spectra of the sample contain strong free– but comparatively weak bound–exciton luminescence implying a high purity of the samples. Tuning the excitation laser frequency into resonance with the free exciton transitions leads to the appearance of additional strong narrow lines with pronounced resonance behaviour (Fig. 9.2a). Their shift with respect to the exciting laser line depends on the laser energy resulting in a strong dispersive behaviour allowing these lines to be interpreted as due to RBS of exciton-polaritons. The clear occurrence of Brillouin scattering means that inelastic scattering by acoustic phonons constitutes the most important mechanism of exciton relaxation and that elastic scattering in k–

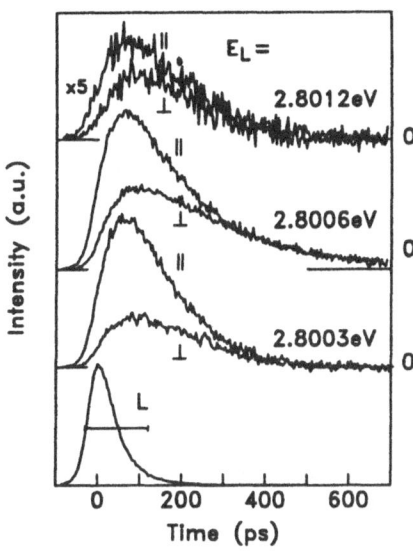

Fig. 9.3
Time dependence of the resonant Brillouin scattered intensity excited at various laser photon energies E_L. The polarization of the laser was parallel to [110], that of the scattered light parallel to [110] ($I_\parallel$) and [1$\bar{1}$0]($I_\perp$). L denotes the time response of the setup to the laser pulse. The time sampling window for measuring the spectra shown in Fig. 9.2 is indicated by the horizontal bar.

space by impurities or defects, which would give rise to resonant Rayleigh scattering, does not blur this process significantly.

Compared to the usual method of Brillouin spectroscopy, by temporal sampling of the scattered intensity only during the laser pulse our technique allows one to efficiently discriminate the fast scattering lines from the longer lived luminescence background as shown in Fig. 9.2b. This has plagued most previous investigations of RBS.

While the energy shifts of the Brillouin lines were analysed in detail in [48, 205] revealing the three polariton branches expected in these samples, here we are interested in the dynamical behaviour of the Brillouin scattering. First results of the polarized decay of the RBS lines excited around the lowest polariton resonance (marked by E_T in the inset to Fig. 9.2) are displayed in Fig. 9.3. As an unpolarized luminescence background has been subtracted, these curves directly reflect the temporal behaviour of the resonant intermediate polariton states. It is clearly seen that the time dependence is nearly independent of excitation energy; only the strength of the interaction varies drastically. At the same time we have found that the polarization degree of the Brillouin scattering lines is only about 20 to 40% . According to the theory in Chap. 5, for RBS the polarization degree of the scattering lines directly reflects the phase relaxation of the intermediate states. Neglecting the transform–limited experimental conditions —the simplest approximation— the measurements can be described assuming a two exponential decay law

$$I_\parallel \propto e^{-t/T_1} + e^{-t/\tau_p} \qquad I_\perp \propto e^{-t/T_1} - e^{-t/\tau_p} \qquad (9.1)$$

with the depopulation time T_1 and the spin relaxation time τ_p which reflects the dephasing of the polariton states. While the depopulation time T_1 is found to be almost energy independent of the order of 150 ps, the spin–relaxation time decreases from about 60 ps at $E_L = 2.800$ eV to 30 ps at 2.802 eV, indicating the increase of dephasing processes for higher exciton energies.

In order to explore fully the exciton–polariton dynamics, in the future these preliminary results have to be extended over the whole range of resonance energies and analyzed in the framework of the transform–limited theory.

9.2 Future Prospects

The present status of the experimental setup allows us to measure the temporal dynamics of light fields down to the time range of about 20 ps with single photon sensitivity. By applying the full power of the transform–limited method, relaxation times down to 5 ps can be deduced reliably by deconvolution procedures. Therefore, one of the most challenging future tasks will be to improve the time resolution of the setup, while still maintaining the high detection sensitivity. This would then allow one to investigate much faster relaxation processes like $LO(\Gamma)$ phonon scattering of excitons or resonance fluorescence from non–stationary wave–packet states in semiconductors which occur on a subpicosecond time scale and up to now could only be investigated by nonlinear techniques (see e.g. [34, 206]).

While the design of appropriate spectral filters with transform–limited response down to the 100 fs time domain can be done quite easily with the design rules given in Sect. 2.3.2, the photodetector obviously is the problematic component. Certainly, the way to improve the time resolution would be to employ streak cameras as detectors. Here at present commercial systems are available that have a time resolution of less than 2 ps [72]. Due to increased gain between the photocathode and the light sensitive two–dimensional image detector (usually a CCD array) these systems now offer single photon detection capability thus increasing the sensitivity and signal–to–noise ratio dramatically [207].

We have adapted our streak camera system (C1587, Hamamatsu Corporation) to reach single photon sensitivity by inserting an image intensifier (TH9304, Thomson) between the phosphor screen and the CCD–full frame image sensor (model TH7883A, Thomson). The output of the CCD is fed at a repetition rate of 10 Hz into a digitizing frame acquisition interface (Data Translation 2856) allowing fast image processing [76]. With the conventional readout using a standard television tube (Grundig FA86 with Si–vidicon 4532), the signal–to–noise ratio and the dynamics of the streak camera/readout system is limited to about two orders of magnitude (compare the lower trace in Fig. 3.7). As shown in Fig. 9.4, after the addition of the image intensifier the sensitivity is high enough to observe the signals from single photoelectrons emitted from the photocathode. Most advantageous, this mode, besides the very high sensitivity, allows one to determine by image

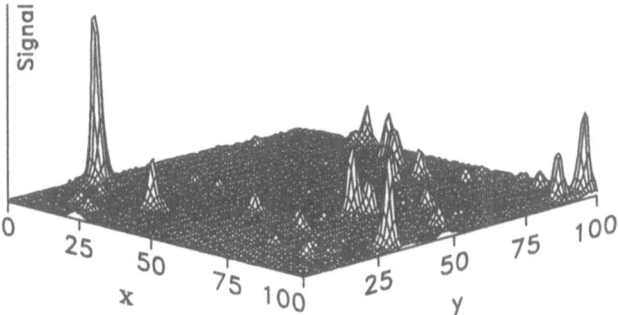

Fig. 9.4 Two–dimensional streak camera image taken with the image–intensified CCD array (x and y denote the array pixels) showing single photon peaks. Note the varying pulse heights of the single photon peaks.

processing the photon arrival time at the cathode with higher resolution. As is shown in Fig. 9.5, a simple amplitude discrimination of the single photon peaks already results in a factor of two improvement of the time resolution of our streak camera system.

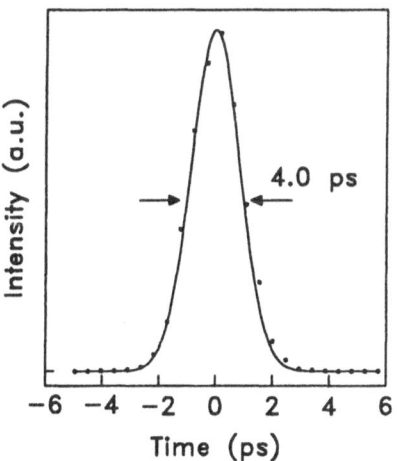

Fig. 9.5
System response to a 2 ps laser pulse measured in the single photon counting mode. From [86].

Using more sophisticated techniques that compensate for the pulse height distribution of the single photon peaks, it can be expected that the ultimate limit in time resolution is given by the timing jitter of the laser and streak camera system. As the latter can be made very small [83] the single–photon mode will offer the possibility of subpicosecond time resolution in the near future.

References

1. Lord Rayleigh (J. W. Strutt), Phil. Mag. **47**, 375 (1899)
2. G. Placzek, in: *Handbuch der Radiologie* **6**, ed. by E. Marx (Akademische Verlagsgesellschaft, Leipzig 1934) p. 209
3. M. Born *Optik* (Springer, Berlin, Heidelberg 1975)
4. W. Heitler *The Quantum Theory of Light* (Clarendon Press, Oxford 1954)
5. R. Loudon *The Quantum Theory of Light*, 2nd edition (Clarendon Press Oxford 1983)
6. D. L. Huber, Phys. Rev. **178**,93 (1969)
7. S. Stenholm, *Foundations of Laser Spectroscopy* (John Wiley & Sons, New York 1984)
8. C. Cohen–Tannoudji, in: *Frontiers in Laser Spectroscopy* 1, ed. by R. Balian, S. Haroche and S. Liberman, (North–Holland Publishing Company, Amsterdam 1977)
9. K. Blum *Density Matrix Theory and Applications* (Plenum Press, New York 1981)
10. M. Aihara, Phys. Rev. **A18**, 606 (1978)
11. L. Knöll and G. Weber, Exper. Technik. Physik, **28**, 299 (1980); **29**, 521 (1981) and **31**, 287 (1983)
12. P. F. Williams, D. L. Rousseau and S. H. Dworetsky, Phys. Rev. Lett. **32**,196 (1974)
13. E. Courtens and A. Szöke, Phys. Rev. **A15**, 1588 (1977)
14. J.H. Eberly and K. Wódkiewicz, J. Opt. Soc. Am. **67**, 1252 (1977)
15. J. Eberly, C. V. Kunasz, and K. Wódkiewicz, J. Phys. B: Atom Molec. Phys. **13**, 217 (1980)
16. T. Takagahara, in: *Relaxation of Elementary Excitations*, ed. by R. Kubo and E. Hanamura (Springer, Berlin, Heidelberg 1980)
17. V. V. Khiznyakov and I. K. Rebane, Sov. Phys. JETP **47**, 463 (1978)
18. I. K. Rebane, A. L. Tuul and V. V. Khiznyakov, Sov. Phys. JETP **50**, 655 (1979)
19. K. Maruyama, F. Shibata and M. Aihara, Physica A **153**, 441 (1988)
20. A. Freiberg and P. Saari, IEEE J. of Quantum Electronics QE19, 622 (1983)
21. P. Saari, J. Aaviksoo, A. Freiberg and K. Timpmann, Opt. Commun. **39**, 94 (1981)
22. H. Stolz and W. von der Osten, Cryst. Lattice Defects Amorphous Mater. **12**, 293 (1985)
23. H. Stolz, V. Langer and W. von der Osten, J. Luminesc. **48**, 72 (1991)

24. H. Stolz, in: *Festkörperprobleme/Advances in Solid State Physics XXXI*, ed. by U. Rössler, (Pergamon, Vieweg, Wiesbaden 1991), p. 219

25. H. Stolz, Phys. Stat. Sol. (b) **173**,99 (1992)

26. R. S. Knox *Theory of Excitons* (Academic Press, New York 1963)

27. L. Schultheiß, A. Honold, J. Kuhl, K. Köhler and C. W. Tu, Phys. Rev. B**34**, 9027 (1986)

28. B. Fluegel, N. Peyghambarian, G. Olbright, M. Lindberg, S. W. Koch, M. Joffre, D. Hulin, A. Migus and A. Antonetti, Phys. Rev. Lett. **59**, 2588 (1987)

29. P. C. Becker, H. L. Fragnito, C. H. Cruz, R. L. Fork, J. E. Cunningham, J. E. Henry and C. V. Shank, Phys. Rev. Lett. **61**, 1647 (1988)

30. J. Kuhl, A. Honold, L. Schultheis and C.W. Tu, in: *Festkörperprobleme/Advances in Solid State Physics IXXX*, ed. by U. Rössler, (Pergamon, Vieweg, Wiesbaden 1989), p. 157

31. E. O. Göbel, K. Leo, T. C. Damen, J. Shah, S. Schmid–Rink, W. Schäfer, J. F. Müller and K. Köhler, Phys. Rev. Lett. **64**, 1801 (1990)

32. B. F. Feuerbacher, J. Kuhl, R. Ecleston and K. Ploog, Sol. State Commun. **74**, 1279 (1990)

33. K. Leo, T. C. Damen, J. Shah, E. O. Göbel and K. Köhler, Appl. Phys. Lett. **57**, 19 (1990)

34. See contribuitons in Proc. NOEKS III Bad Honneff 1993, Phys. Stat. Sol (b) **173**, 1 (1992)

35. W. Kaiser (ed.) *Ultrashort Laser Pulses and Applications*, Topics in Applied Physics, **60**, (Springer, Berlin, Heidelberg 1988)

36. M. Cardona and G. Güntherodt (eds.): *Light Scattering in Solids II – VI*, (Springer, Berlin, Heidelberg 1982 – 1991)

37. J. Aaviksoo, A. Freiberg, T. Reinot and S. Savikhin J. Luminesc. **35**, 267 (1986) and J. Aaviksoo, A. Freiberg, J. Lippmaa and T. Reinot J. Luminesc. **37**, 267 (1987)

38. J. Aaviksoo, J. Lumin. **48 & 49**, 57 (1991)

39. A. Z. Genack, H. Z. Cummins, M. A. Washington and A. Compaan, Phys. Rev. B**12**, 2478 (1975)

40. P. Y. Yu and Y. R. Shen, Phys. Rev. B**12**, 1377 (1975)

41. J. Windscheif and W. von der Osten, J. Phys. C **13**, 6299 (1980)

42. W. von der Osten, in: *Light Scattering in Solids VI*, Topics Appl. Phys. **68**, ed. by M. Cardona and G. Güntherodt (Springer, Berlin, Heidelberg 1991) p. 361

43. J. Weber and W. von der Osten, Z. Phys. B**24**, 343, (1976)

44. H. Stolz and W. von der Osten, Solid State Commun. **49**, 1035 (1984)

45. H. Stolz, E. Schreiber and W. von der Osten, in: *Proc. 17th Int. Conf. Phys. Semiconductors*, San Francisco, USA, 1984, ed. by J. D. Chadi and W. A. Harrison (Springer, New York 1985) p. 1271

46. J. S. Weiner and P. Y. Yu, Solid State Commun. **50**, 493 (1984)

47. Th. Weber, H. Stolz, W. von der Osten, Zhao Futan and Shen Yongrong, Superlattices and Microstruct. **13**, 359 (1993)

48. S. Permogorov, H. Vogelsang, Th. Weber, H. Stolz, W. von der Osten, P. Kuznetov, A. N. Pechonov and A. S. Nasibov, Solid State Commun. **88**, 705 (1993)

49. M. V. Klein, Phys. Rev. B8, 919 (1973)
50. Y. R. Shen, Phys. Rev. B9, 622 (1974)
51. J. R. Solin and H. Merkelo, Phys. Rev. B12, 624 (1975); Phys. Rev. B14, 1775 (1976)
52. H. Stolz, D. Schwarze, W. von der Osten and G. Weimann, Phys. Rev. B47, 9669 (1993)
53. A. Vanettieri, J. Shah, T. C. Damen, D. S. Kim, L. N. Pfeiffer and L. J. Sham, Journ. de Phys. IV, C5 3, 27, (1993)
54. H. M. Nussenzweig *Introduction to Quantum Optics* (Gordon and Breach, London, New York 1973)
55. R. L. Fork, C. M. Brito Cruz, P. C. Becker and C. V. Shank,Opt. Lett. 12, 483 (1987)
56. A. Takahashi, M. Nishizawa, Y. Inagaki, M. Koishi, and K. Kinoshita, Proc. Soc. Photo-Opt. Instrum. Eng. 2116
57. J. Shah, IEEE J. of Quantum Electronics QE24, 276 (1988)
58. C. H. Page, J. Appl. Phys. 23, 103 (1952)
59. D. G. Lampard,J. Appl. Phys. 25, 803 (1954)
60. H. E. Ponath and M. Schubert, Annal. Physik 37, 109 (1980)
61. H. Wegener, *Der Mössbauereffekt und seine Anwendung in Physik und Chemie* (Bibliographisches Institut, Mannheim 1966)
62. K.-H. Brenner and K. Wódkiewicz, Optics Commun. 43, 103 (1982)
63. H. Stark, *Image Recovery: Theory and Applications*, Academic, Orlando (1987)
64. D. J. Kane and R. Trebino, IEEE J. of Quantum Electronics QE29, 571 (1993)
65. W. Demtröder *Laser Spectroscopy* Springer Series in Chemical Physics 5 (Springer, Berlin, Heidelberg 1991)
66. M. Abramowitz and I. A. Stegun *Handbook of Mathematical Functions* (Dover, New York 1965)
67. H. Noda, T. Namioka and M. Seya,J. Opt. Soc. 64, 1031 (1974); J. Opt. Soc. 64, 1037 (1974)
68. R. Gase and M. Schubert, Optica Acta 29, 109 (1982) and Optica Acta 30, 1125 (1983)
69. H. Kalt, W. W. Rühle, K. Reimann, M. Rinker and E. Bauser, Phys. Rev. B43, 12364 (1991)
70. N. H. Schiller, Y. Tsuchiya, E. Inuzuka, Y. Suzuki, K. Kinoshita, H. Lida and R. R. Alfano, Optical Spectra 14, 55 (1980)
71. Manual of streak camera C1587 (Hamamatsu Corporation, Hamamatsu, Japan)
72. Data sheet C5680, Hamamatsu Corporation, Hamamatsu City (1993)
73. E. Schreiber, Ph. D. thesis, Universität Paderborn (1989)
74. D. Schwarze, Ph. D. thesis, Universität Paderborn (1991)
75. M. Kiene, Diploma thesis, Universität Paderborn (1992)
76. V. Langer, Ph. D. thesis, Universität Paderborn (1992)
77. C. Sommer, Diploma thesis, Universität Paderborn (1991)
78. F. Krausz, Ch. Spielmann, T. Brabec, E. Wintner, and A. J. Schmidt, Optics Letters 17, 204 (1992) and references therein

79. E. Reed and G. Franginas, in: *Solid State Lasers*, ed. by G. Dube, Proc. Soc. Photo-Opt. Instrum. Eng. **1223**

80. B. Schicht, private communication

81. J. Herrmann and B. Wilhelmi *Laser für ultrakurze Lichtimpulse* (Akademie–Verlag, Berlin 1984) and
    K. Ibbs, Lasers & Optronics, **1990**, p. 29

82. E. Schreiber, Diploma thesis, Universität Paderborn (1984)

83. A. Finch, W. E. Sleat and W. Sibbett, Rev. Sci. Instrum. **60**, 839 (1989)

84. M. B. Morris and T. J. McIlrath, Applied Optics **18**, 345 (1979)

85. Th. Weber, Diploma thesis, Universität Paderborn (1989)

86. B. Schicht, Diploma thesis, Universität Paderborn, Paderborn, (1994)

87. see e.g. Ref. [65]

88. Operators manual TC 455, Tennelec Corporation, Oak Ridge, USA (1988)

89. A. Elsner, Diploma thesis, Universität Paderborn (1988)

90. H. Stolz, V. Langer, E. Schreiber S. Permogorov and W. von der Osten, Phys. Rev. Lett. **67**, 679 (1991)

91. D. Sommer, Diploma thesis, Universität Paderborn (1984)

92. E. I. Rashba and M. D. Sturge (eds.) *Excitons* (North–Holland Publishing Comp. 1982)

93. S. Koch and H. Haug *Quantum Theory of the Optical and Electronic Properties of Semiconductors*, 2nd edition (World Scientific, Singapore 1993)

94. G. L. Bir and G. E. Pikus *Symmetry and Strain–induced Effects in Semiconductors* (Keter Publishing House, Jerusalem 1974)

95. H. Haken, in: *Proceedings of the International School of Physics "Enrico Fermi", Course 64*, ed. by N. Bloembergen Varenna, Italy (1977)

96. J. Röseler, Habilitation thesis, Humboldt Universität Berlin (1985)

97. K. Seeger, *Semiconductor Physics* (Springer, New York, Wien 1973)

98. Y. Toyozawa, Prog. Theor. Phys. **20**, 53 (1958)

99. F. Bassani and G. Pastori Parravicini *Electronic States and Optical Transitions in Solids* (Pergamon Press, Oxford 1975)

100. G. F. Koster, J. O. Dimmock, R. G. Wheeler, H. Statz, *Properties of the Thirty-Two Point Groups* (M.I.T. Press, Cambridge Massachusetts 1963)

101. H. Haken and A. Schenzle in: *Polaritons, Proceedings of the First Taormina Research Conference on the Structure of Matter*, ed. by E. Burstein and F. de Martini, Taormina, Italy (1973), (Pergamon Press, New York 1974)

102. F. Bassani and L. C. Andreoni, in: *Proceedings of the International School of Physics "Enrico Fermi", Course 96*, ed. by U. M. Grassano and N. Terzi, Varenna, Italy (1985)

103. A. Stahl and I. Balslev *Electrodynamics of the Semiconductor Band Edge*, Springer Tracts in Modern Physics **110** (Springer, Berlin, Heidelberg 1987)

104. S. Schmitt–Rink, D. S. Chemla and H. Haug, Phys. Rev. **B37**, 941 (1988)

105. I. Balslev, R. Zimmermann and A. Stahl, Phys. Rev. **B40**, 4095 (1989)

106. W. Greiner, *Theoretische Physik* **4a** (Harri Deutsch, Thun und Franfurt a. Main 1989)

107. R.J. Elliott, Phys. Rev. **124**, 340 (1961)

108. A. Quattropani, L. C. Andreani and F. Bassani, Il Nuovo Cim. D **7**, 55 (1986); F. Bassani, F. Ruggiero and A. Quattropani, Il Nuovo Cim. D **7**, 700 (1986)

109. A. S. Dawydow, *Quanten Mechanik*, 7th edition, (VEB Verlag, Berlin 1987)
110. R. H. Lemberg, Phys. Rev. A2, 883 (1969)
111. J.J. Hopfield, Phys. Rev. 112, 1555 (1958)
112. V. M. Agranovich and V. L. Ginzburg, in: *Crystal Optics with Spatial Dispersion and Excitons*, Springer Series in Solid State Sciences 42, ed. by V. M. Agranovich and V. L. Ginzburg (Springer, Berlin, Heidelberg 1984)
113. C. Weisbuch and R. G. Ulbrich, in *Light Scattering in Solids III*, eds. M. Cardona, G. Güntherodt, Springer, Berlin, Heidelberg (1982), p. 207
114. M. Artoni and J. L. Birman, Phys. Rev. B44, 3736 (1991)
115. D. Fröhlich, A. Kulik, B. Uebbing, A. Mysyrowicz, V. Langer, H. Stolz, and W. von der Osten, Phys. Rev. Lett. 67,2343 (1991)
116. U. Sliwczuk, Ph. D. Thesis, Universität Paderborn (1987); U. Sliwczuk, H. Stolz and W. von der Osten, phys. stat. sol. (b) 122, 203 (1984)
117. R. Zimmermann, phys. stat. sol. (b) 173, 129 (1992)
118. H. Haug, phys. stat. sol. (b) 173, 139 (1992)
119. K. H. Ploog and L. Tapfer (Eds.), *Physics and Technology of Semiconductor Quantum Devices*, Lecture Notes in Physics 419 (Springer, Berlin, Heidelberg 1993) and C. Weisbuch and B. Vinter, *Quantum Semiconductor Structures: Fundamentals and Applications* (Academic Press, New York 1991)
120. E. O. Göbel and K. Ploog, Prog. Quant. Electr. 14, 289 (1990)
121. G. Bastard, *Wave Mechanics Applied to Semiconductor Heterostructures* (les éditions de physique, Paris 1990)
122. M. Altarelli, Phys. Rev. B28, 842 (1983)
123. E. O. Kane, Phys. Rev. B11, 3850 (1975)
124. A. I. Ansel'm and Y. A. Firsov, Soviet Phys. JETP 1, 139 (1955); ibid. 3, 564 (1956)
125. P. Y. Yu, in: *Excitons*, ed. by K. Cho, Topics in Current Physics 14 (Springer, Berlin, Heidelberg 1979) p. 211
126. R. Zimmermann, private communication
127. T. Takagahara, Phys. Rev. B31, 65552 (1985)
128. J. Lee, E. S. Koteles and M. O. Vassel, Phys. Rev. B33, 5512 (1986)
129. H. Stolz, D. Schwarze, W. von der Osten and G. Weimann, Superlattices and Microstruct. 6, 271 (1989)
130. G. E. Pikus and E. L. Ivchenko, in: *Excitons*, ed. by E. I. Rashba and M.D. Sturge (North–Holland Publishing Comp. 1982)
131. B. Crosignani, P. Di Porto and M. Bertolotti *Statistical Properties of Scattered Light* (Academic Press, New York 1975)
132. J. Hegarty, M. D. Sturge, C. Weisbuch, A.C. Gossard and W. Wiegmann, Phys. Rev. Lett., 49, 930 (1982)
133. J. Hegarty, L. Goldner, and M.D. Sturge, Phys. Rev. B30, 7346 (1984)
134. J. Hegarty and M.D. Sturge, J. Opt. Soc. Am. B 2, 1143 (1985)
135. A.A. Maradudin and D.L. Mills, Phys. Rev. B11, 1392 (1975)
136. V.A. Kosobukin and A.V. Sel'kin, Solid State Commun. 66, 313 (1988)
137. V.A. Kosobukin, M.I. Sazhin, and A.V. Sel'kin, Sov. Phys. Solid State 32 (1990)
138. M. Born and E. Wolf *Principles of Optics* (Pergamon, Oxford 1970)
139. J. Aaviksoo and J. Kuhl, IEEE J. Quant. Electr. 25, 2523 (1989)

140. H. Stolz, D. Schwarze, W. von der Osten and G. Weimann, Superlattices and Microstruct. **9**, 511 (1991)

141. M. Aihara, Phys. Rev. Lett. **57**, 463 (1986)

142. S. Haroche, in: *High Resolution Laser Spectroscopy*, ed. by K. Shimoda (Springer, Berlin, Heidelberg 1976)

143. W. Harper, Rev. Mod. Phys. **44**, 169 (1972)

144. M. Aihara and A. Kotani, Solid State Commun. **46**, 751 (1983)

145. Y. Toyozawa, J. Phys. Soc. Japan **41**, 400 (1975)

146. A. Kotani and Y. Toyozawa, J. Phys. Soc. Japan **41**,1699 (1976)

147. E. Fick *Einführung in die Grundlagen der Quantentheorie* 2. edition (Akademische Verlagsgesellschaft, Frankfurt am Main 1972)

148. E. L. Ivchenko, I. G. Lang and S. T. Pavlov, Sov. Phys. Solid State **19**, 1610 (1977)

149. A. Nakamura, M. Shimura, M. Hirai, M. Aihara and S. Nakashima, Phys. Rev. B**35**, 1281 (1987)

150. V. Langer, H. Stolz and W. von der Osten, Phys. Rev. Lett. **64**, 854 (1990)

151. V. Langer, H. Stolz and W. Von der Osten, J. Luminesc. **45**, 406 (1990)

152. F. Bassani, R. S. Knox and W. B. Fowler, Phys. Rev. **129**, 2554 (1965)

153. P. M. Scop, Phys. Rev. **139**, A934 (1963)

154. H. Overhof in: *I–VII Compounds, Landolt–Börnstein, Numerical Data and Functional Relationships in Science and Technology* New Series III 17b *Semiconductors*, ed. by O. Madelung, (Springer, Berlin Heidelberg 1982); 22a, (1987)

155. B. L. Joesten and F. C. Brown, Phys. Rev. **148**, 919 (1969)

156. G. Ascarelli, Phys. Rev. **179**, 797 (1969)

157. K. Cho, Phys. Rev. B**14**, 4463 (1976)

158. O. Madelung (ed.) *I–VII Compounds, Landolt–Börnstein, Numerical Data and Functional Relationships in Science and Technology* New Series III 17b *Semiconductors*, (Springer, Berlin, Heidelberg 1982)

159. H. Stolz, W. Wassmuth, W. von der Osten and Ch. Uihlein, J. Phys. C **16**, 955 (1983)

160. M. Matsushita,J. Phys. Soc. Japan **35**, 1688 (1973)

161. B. Dorner, W. von der Osten and W. Bührer, J. Phys. C **9**, 723 (1976)

162. R. M. A. Azzam and N. M. Bashara *Ellipsometry and Polarized Light* (North-Holland Publishing Co., Amsterdam 1977)

163. V. Langer, private communication

164. V. Langer, H. Stolz, W. von der Osten, D. Fröhlich, A. Kulik, and B. Uebbing, Europhys. Lett., **18**, 723 (1992)

165. V. Langer, H. Stolz, and W. von der Osten, submitted to Phys. Rev B

166. V.T. Agekyan, phys. stat. sol. (a) **43**, 11 (1977)

167. D. Fröhlich and R. Kenklies, phys. stat. sol. (b) **111**,247 (1982)

168. G. Kubawara, M. Tanaka and H. Fukutani, Solid State Commun. **21**, 599 (1977)

169. D. Fröhlich, A. Kulik, B. Uebbing, V. Langer, H. Stolz and W. von der Osten, Phys. Stat. Sol (b) **173**, 31 (1992)

170. A. Kulik, Diploma thesis, Universität Dortmund (1992)

171. J. L. Birman and R. Berenson, Phys. Rev. B9, 4512 (1974) and J. L. Birman, Phys. Rev. B9, 4518 (1974)

172. G. Moruzzi and F. Strumia (Eds.) *The Hanle Effect and Level Crossing Spectroscopy* (Plenum, New York 1991)

173. B. Bendow, Springer Tracts in Modern Physics, 82, (Springer, Berlin, Heidelberg 1978) p. 89.

174. S.I. Pekar, Sov. Phys. JETP 6, 758 (1958)

175. H.-R. Trebin, H.Z. Cummins and J.L. Birman, Phys. Rev. B23, 597 (1981)

176. M.H. Manghnani, W.S. Brower and H.S. Parker, phys. stat. sol. (a) 25, 69 (1974)

177. I–VII Compounds, Landolt–Börnstein, Numerical Data and Functional Relationships in Science and Technology New Series III 17b Semiconductors O. Madelung (Hrsg.), Springer, Berlin (1982)

178. M. Jütte, Diploma thesis, Universität Paderborn (1992)

179. H. Mathieu, Y. Chen, J. Camassel, J. Allegre, and D. S. Robertson, Phys. Rev. B32, 4042 (1985)

180. J. Gutowski, N. Presser and G. Kudlek, phys. stat. sol. (a) 120, 11 (1990)

181. O. Madelung (ed.) *III–V Compounds, Landolt–Börnstein, Numerical Data and Functional Relationships in Science and Technology* New Series III 17a *Semiconductors* (Springer, Berlin, Heidelberg 1982)

182. S. Permogorov and A. Reznitsky, Reprint 1555, Ioffe–Institut, St. Petersburg; J. Luminesc. to be published

183. W. von der Osten and H. Stolz, J. Phys. Chem. Solids 51, 765 (1991)

184. N. F. Mott and E. A. Davies, *Electronic Properties in Non–Crystalline Materials* (Clarendon Press, Oxford 1971)

185. E. N. Economou, C. M. Soukoulis, M. H. Cohen and S. John, in: *Disordered Semiconductors*, ed. by M. A. Kastner, G. A. Thomas and S. R. Ovskinsky (Plenum, New York 1987) p. 681

186. U. Siegner, D. Weber, E. O. Göbel, D. Bennhardt, V. Heuckeroth, R. Saleh, S. D. Baranovskil, P. Thomas, H. Schwab, C. Klingshirn, J. V. Hvam and V. G. Lyssenko,Phys. Rev B 46, 4564 (1992)

187. S. Jorda, Journ. de Phys. IV, C5 3, 59, (1993)

188. M. Heuken, K. P. Geyzers, J. Söllner, A. Schneider, F. E. G. Guimarães, and K. Heime, J. Crystal Growth, 124, 633 (1992)

189. J. Christen and D. Bimberg, Phys. Rev. B42, 7213, (1990)

190. B. Sermage, S. Long, B. Devaud, and D. S. Katzer, Journ. de Phys. IV, C5 3, 19, (1993)

191. D. Bennhardt, P. Thomas, A. Weller, M. Lindbergh and S. W. Koch, Phys. Rev. B43, 8934 (1991)

192. J. Feldmann, G. Peter, E. O. Göbel, P. Dawson, K. Moore, C. Foxon and R. J. Elliott, Phys. Rev. Lett. 59, 2337 (1987)

193. E.I. Rashba and G.E. Gurgenishvili, Fizika Tverdogo Tela 4, 1029 (1962), engl. transl.: Sov. Phys.–Solid State 4, 759 (1962)

194. R. Eccleston, B. F. Feuerbacher, J. Kuhl, W. W. Rühle and K. Ploog, Phys. Rev. B45, 5555 (1992)

195. K. T. Tsen, O. F. Sankey, and H. Morkoç, Appl. Phys. Lett. 57, 1666 (1990)

196. D. Overhauser, K.-H. Pantke, J. M. Hvam, G. Weimann, and C. Klingshirn, Phys. Rev. B47, 6827 (1993)
197. M. Z. Maialle, E. A. de Andrada e Silva, and L. J. Sham, Phys. Rev. B47, 15776 (1993)
198. V. Langer and D. Schwarze, unpublished measurements
199. L. C. Andreani, Physica Scripta T35, 115, (1991)
200. D. S. Citrin, Phys. Rev. B47, 38332 (1993)
201. D.G. Thomas and J. J. Hopfield, Phys. Rev. 128, 2135 (1962)
202. O. Madelung (ed.) *II–VI Compounds, Landolt–Börnstein, Numerical Data and Functional Relationships in Science and Technology* New Series III 17b *Semiconductors* Springer, Berlin, Heidelberg (1982)
203. W. Breing, R. Zeyher, and J. Birman, Phys. Rev. B6, 4617 (1972)
204. R. G. Ulbrich and C. Weisbuch, Phys. Rev. Lett. 38, 865 (1977)
205. H. Mayer, U. Rössler, S. Permogorov, H. Stolz, H. Vogelsang, and W. von der Osten, Proc. Sixth International Conference on II–VI Compounds, Newport, Rhode Island (USA) 1993
206. J. Feldmann, T. Meier, G. von Plessen, M. Koch, E. O. Göbel, P. Thomas, G. Bacher, C. Hartmann, H. Schweizer, W. Schäfer, and H. Nickel, Phys. Rev. Lett. 70, 3027 (1993)
207. L. M. Davis and C. Parigger, Meas. Sci. Technol. 3, 85 (1992)

Subject Index

Springer Tracts in Modern Physics

---

* denotes a volume which contains a Classified Index starting from Volume 36

Springer-Verlag
and the Environment

We at Springer-Verlag firmly believe that an international science publisher has a special obligation to the environment, and our corporate policies consistently reflect this conviction.

We also expect our business partners – paper mills, printers, packaging manufacturers, etc. – to commit themselves to using environmentally friendly materials and production processes.

The paper in this book is made from low- or no-chlorine pulp and is acid free, in conformance with international standards for paper permanency.